Contents

Mathematical Illustrations

A MANUAL OF GEOMETRY AND POSTSCRIPT

BILL CASSELMAN

University of British Columbia

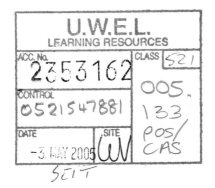

CAMBRIDGE
UNIVERSITY PRESS

PUBLISHED BY THE PRESS SYNDICATE OF THE UNIVERSITY OF CAMBRIDGE
The Pitt Building, Trumpington Street, Cambridge, United Kingdom

CAMBRIDGE UNIVERSITY PRESS
The Edinburgh Building, Cambridge CB2 2RU, UK
40 West 20th Street, New York, NY 10011-4211, USA
477 Williamstown Road, Port Melbourne, VIC 3207, Australia
Ruiz de Alarcón 13, 28014 Madrid, Spain
Dock House, The Waterfront, Cape Town 8001, South Africa

http://www.cambridge.org

First published 2005

Printed in Hong Kong, China

Typefaces Photina MT 10/13.5 pt. with ITC Symbol and Lucida Sans Typewriter
System $\LaTeX\,2_\varepsilon$ [TB]

A catalog record for this book is available from the British Library.

Library of Congress Cataloging in Publication Data
Casselman, Bill, 1941–
    Mathematical illustrations : a manual of geometry and PostScript / Bill Casselman.
        p.   cm.
    Includes bibliographical references and index.
    ISBN 0-521-83921-1 (hardback) – ISBN 0-521-54788-1 (pbk.)
    1. PostScript (Computer program language)   I. Title.
  QA76.73.P67C37   2004
  005.13′3 – dc22   2004045886

ISBN 0 521 83921 1 hardback
ISBN 0 521 54788 1 paperback

PostScript®, Illustrator®, and PhotoShop® are registered trademarks of Adobe Systems,
Inc. *Mathematica*® is a registered trademark of Wolfram Research. Maple™ is a
trademark of Waterloo Maple Inc. MATLAB® is a registered trademark of the Mat
Works, Inc. Windows® is a registered trademark of Microsoft Corporation. Macintosh®
is a registered trademark of Apple Computers, Inc. UNIX® is a registered trademark of
The Open Group in the United States and other countries. Other proprietary names used in
the book are registered trademarks, and where possible, this is indicated within the text.

In this book, the phrase, "PostScript interpreter" mean an interpreter of the PostScript language.

The specific list of commands that make up the PostScript language is copyrighted by
Adobe Systems, Inc.

Preface

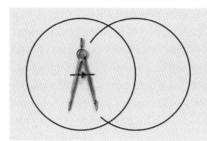

```
1 0 0 setrgbcolor
newpath
0 0 1 0 360 arc
stroke
newpath
1 0 1 0 360 arc
stroke
```

This book will show how to use PostScript to produce mathematical graphics at several levels of sophistication. It also includes some discussion of the mathematics involved in computer graphics as well as a few remarks about good style in mathematical illustration. It has been used as a textbook in an undergraduate course in geometry.

Nowadays, of course, there are many tools to help in this task. A partial list would include the free packages xfig, pictex, PSTricks, MetaFont, and MetaPost as well as commercial mathematical programs such as Maple and Mathematica® and professional graphics design tools such as Illustrator®. Which one to choose apparently involves a trade-off between simplicity and quality in which most illustrators opt for whatever is perceived to be simplicity. The truth is that the trade-off is unnecessary; once you have made a small initial investment of effort, by far the best thing to do in most situations is to write a program in the graphics programming language PostScript. There is practically no limit to the quality of a PostScript program's output, and as one acquires experience the difficulties of using the language decrease rapidly. The apparent complexity involved in producing simple figures by programming in PostScript, as I hope this book will demonstrate, is largely an illusion. And the amount of work involved in producing more complicated figures will usually be neither more nor less than is necessary.

PostScript is a rather low-level computer language developed by Adobe Systems in its startup days. Its primary purpose is to produce high-quality graphics from computers and especially to output computer graphics on printers. Professional graphics designers, for example, often work with high-end programs that in turn produce their output through PostScript. To them, the PostScript itself is usually invisible. One would not therefore expect PostScript to be comprehensible to humans. It is nonetheless a convenient computer language for producing by hand, so to speak, designs with a significant mathematical component. One great virtue of PostScript is that there is no effective limit on the *quality* of the final product precisely because it is designed for use by professionals. Some other virtues are that it is amenable to "programming by cut and paste" and that the effects of commands are very localized – local, that is, within the text of the program. Both virtues and drawbacks will become better known as we proceed.

PostScript allows essentially complete control over the final product, which is something impossible with all of the graphics packages I listed above. Having such fine control over your figures means that, once your code is in place, it is often quite easy to modify it. This makes it a great tool for discovering, not only explaining, mathematical ideas. The advantage of this control will be very evident to those who have used xfig or pictex, for instance, and less perhaps to users of Maple and Mathematica. What becomes apparent in the course of heavy usage, however, is that a program like Maple is designed for graphics only incidentally. It produces *huge* files – really, unnecessarily huge – and in practice seems reluctant to draw exactly what you want to draw. It can do all sorts of interesting computations, but normally the best way to use this talent is to have it output data files that a PostScript program can then access. The program Mathematica seems to be better adapted for graphics, but there are still many simple tasks it has trouble with. In addition, all Mathematica figures somehow preserve in them the rather strong flavor of Mathematica, whereas PostScript is far more neutral and is better able to take on a characteristic quality of your own devising. This can be a source of great satisfaction.

One competitor to PostScript is the programming language MetaPost. It is an extension of Donald Knuth's MetaFont written by John Hobby, once a graduate student of Knuth's and the developer of the marvelous utility dvips. The programming language itself is basically MetaFont, but the output is to PostScript rather than directly to a physical device. MetaFont is certainly an interesting language – one nicely adapted if little used for font design. There are a few well-known mathematicians who are very fond of it. It has many fine features, but I for one find it much less intuitive to use than PostScript itself as an all-purpose graphics language.

One big advantage of using PostScript is that a PostScript interpreter of good quality exists that costs nothing and runs on nearly all platforms – the program Ghostscript, which I'll refer to throughout this book. Appendix 2 explains how to install it on your computer if it is not already there and how to configure it for a comfortable working environment.

The principal disadvantage of using PostScript is that, for complicated computations, it is not as efficient either to program or to run as the more standard languages such as Java or C. This problem has an easy solution: with just a little knowledge of PostScript you can write a program in one of those languages that will produce a PostScript file.

There are some other disadvantages of PostScript, however. Anyone who takes up this book seriously will realize this quite quickly, and so I had better stave off disappointment.

For one thing, PostScript is not a language with much structure. This, however, turns out to be both a plus and a minus – a plus because you can program in PostScript by "cut and paste" without worrying too much about the consequences. For another, the complete language possesses an intimidating number of commands, but in practice the basic drawing language is rather simple. Almost anyone familiar with coordinate geometry can learn how to do some interesting drawing with the language in a remarkably short time. For more complicated work one will want to build up a library of useful procedures. This book will explain a few such libraries, and in particular one that extends PostScript to three dimensions.

Another bad feature of the language is that debugging and error handling are both atrocious – at least with the interpreters available today. It would be nice if someone would come up with an interpreter that could trace a PostScript program line by line while keeping an eye on its internal state. Such a tool would make development a far easier job. Making an interpreter like this doesn't look to be too difficult a task if the objective is only to establish a good environment for technical graphics (as opposed to commercial production). For this relatively restricted task it wouldn't be necessary to interpret the whole language, and in fact it wouldn't even be absolutely necessary to interpret PostScript itself but merely something close to it. But in any event such an interpreter hasn't been built yet.

And, finally, the integration of text and graphics, although quite feasible, is not as easy as it might be. There seems to be no graphical problem more vexing to those who make mathematical diagrams than that of how to place TEX labels in a diagram. This book will explain (in Appendix 7) how to do it, if awkwardly, in a PostScript figure. Again, what is needed is a tool that doesn't exist but that shouldn't be too hard to produce – one that creates technical text displays that can be imported easily into a basic background diagram and preferably has a simple mouse interface for text editing, scaling, and placement. In the end, such a program presumably would merely import a PostScript program produced by TEX into another PostScript file. There used to be a demonstration program named Import, made available by Adobe Systems when Display PostScript was first distributed on computers running X-Windows, that was better than anything else now available. It didn't do much, but it was refreshingly simple. Among its virtues was that its output was basically no more complicated than it had to be, which meant that you didn't have to stop working on a figure once you had "imported" it. Commercial graphics programs (such as Adobe's Illustrator) are capable of importing PostScript files, but they are expensive. In addition they will take your PostScript programs up a one-way street – what they produce will be far more complicated than what you started with. They will mangle all your nice handwritten code, and the output will be impossible to adjust subsequently. Alas, `Import` was taken seriously by only a few, was always prone to bugs, ran on only a few platforms, and has disappeared.

Even with all its drawbacks, however, PostScript is the tool of choice for making good-quality mathematical illustrations. This book should therefore be of interest to a large group of people: (1) to any scientist who has to make up a fair number of technical illustrations but has been limited by what the other packages can do; (2) to any teacher, even at the secondary school level, who would like to make mathematics a more comprehensible and better motivated subject to his students; (3) to students who want to see how even elementary mathematics that might have seemed useless when first encountered can be applied to attractive real-world problems; and (4) even to artists who might be interested in abstract designs with a mathematical component.

Despite the difficulties of programming in PostScript, it is overall an enjoyable language to work with largely because the final product can be truly beautiful. Post Script is above all a complete graphics language, the only real limitation in working with it is ultimately one's imagination. It is capable of extreme subtlety, brilliant color, and infinite variety. What more does one want? But I must recall a warning, which, although written in an earlier time, is still valid today:

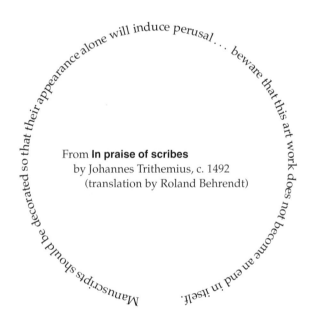

...appearance alone will induce perusal... beware that this art work does not become an end in itself. Manuscripts should be decorated so that their

From **In praise of scribes**
by Johannes Trithemius, c. 1492
(translation by Roland Behrendt)

USING THIS BOOK AS A COURSE TEXT

Perhaps the most surprising feature of this book is that it has been used successfully as a text for a third-year undergraduate course in geometry. This is not at all an obvious use for it, and I think it is fair to say that many readers will become more puzzled by this claim in skimming the book. It is certainly not an orthodox mathematics textbook, nor is it an orthodox text about computer graphics. As one person commented, it is perhaps best described as a text in *metamathematics*, and it is true that the book does cover some basic results in geometry; however, what it really does is provide a tool for students to use themselves in exploring many different fields of mathematics. The book has in fact a mildly recursive quality, for it might be briefly described as showing the reader how to produce a book about geometry!

There were two ways in which the book functioned well in the role of course text. The first is that students find drawing in PostScript to be a definite source of pleasure once an initial break-in period has passed. For this, however, it is necessary to acquire an intimate knowledge of coordinate geometry, linear algebra, and – as various chapters of this book demonstrate – projective and differential geometry. In other words, the pleasures of drawing serve to motivate a great deal of mathematics that many students otherwise find it difficult to become interested in. To some extent, it might be called a text in *applied geometry*.

A second way is more subtle. Throughout the one-term course in which I use this book as a text, students are continually asked to interpret classical mathematics – for example, Euclid's **Elements**, Archimedes' works, and Newton's **Principia** – in illustrations done at first by careful hand sketching and later with PostScript. For this, they are required to go to the original sources (in translation). Reading mathematical classics is ordinarily considered difficult, and this is the best way I know of to get even weak students to digest the material rather than just acquire it formally.

> *In rendering mathematics in images, a person is forced to understand the essentials of an argument rather than to get lost in details.*

This art, if I may call it that, of seeing the essential parts of mathematical proofs is not easy to teach in the traditional course.

In the one-term course I teach, there is not time enough to cover the whole book. Instead I go through the first five chapters in roughly five weeks, including excursions into Euclid, and then spend the remainder of the term in three dimensions (so to speak). Much of the course taught in this way has served well for students of computer science simultaneously taking a practical course in computer graphics – a course in which they usually learn few of the reasons for what they are doing. I might add that a small but intriguing number of students every year – perhaps two or three out of thirty – exhibit an impressive mathematical talent that would not likely be seen in the traditional curriculum. In any event, a major component of the course consists of a student project in which students are required to explain a mathematical topic of their own choosing by means of illustrations done with PostScript. Some of these can be seen at

http://www.math.ubc.ca/~cass/courses/student-projects/

This site also shows some projects from a sequel course on conic sections. An ideal way to use the material from this book in teaching is to spend one term on material from the book itself followed by one term on a more traditional subject in which many of the exercises are required to be done as illustrations.

There is one feature of the course I teach that draws many questions. Given that there are numerous ways in which to use computers in a geometry course, why do I teach the use of PostScript? There are several reasons. One is that its very low-level nature forces students to solve mathematical problems in order to draw. Another is that a student using it doesn't have the feeling that there is something going on that could not be understood if he or she looked into it carefully. A third is that, although the language itself is very low level, as the course develops the student acquires several libraries that extend it considerably. I would expect that anyone using the book for a course would inevitably make up such collections of procedures too in order to deal with problems specific to that particular course. The instructor and student are, for better or worse, likely both required to be active rather than passive.

A FINAL REMARK

It is not as widely acknowledged as it might be that to explain mathematical ideas well often requires good illustrations, and computers in our age have drastically changed the potential of graphical output for this purpose. There are many aspects to this change. The most apparent is that computers allow the user to produce a volume of graphics output never before imagined. A less obvious one is that computers have made it possible for amateurs to produce their own illustrations of professional quality – possible, but not easy, and certainly not as easy as it is to produce their own mathematical text with Donald Knuth's program TEX. In spite of the advances in technology over the past 50 years, it is still not a trivial matter to come up routinely with figures that show exactly what you want them to show exactly where you want to show it. This is to some extent inevitable. Pictures at their best contain much information, and by definition this means that they are capable of unpredictable variety. It is surely not possible to come up with a really simple tool that will let you create easily all the graphics you want to create – the range of possibilities is just too large. All you can hope for is that the amount of work involved in producing an illustration is in proportion to the intrinsic difficulty of what you want to do. And the intrinsic difficulty of producing a good mathematical illustration means that you should expect to do some interesting mathematics as well as to solve interesting computational problems along the way. Mathematical illustrations are a special breed. A really good mathematical illustration almost always requires mathematics in the process of making it.

OUTLINE

The early chapters (1, 3–6) offer an introduction to basic features of PostScript. Reading these chapters alone will get a reader a long way to coming up with good illustrations. Chapters 2 and 12 offer accounts of the basic mathematics involved in mathematical graphics computation – mostly coordinate geometry in 2D and 3D. Chapters 7–10 explore more sophisticated features of PostScript in 2D as well as how mathematics and graphics algorithms interact in interesting ways. Chapters 11, 13, and 14 explore the problems of drawing in three dimensions using a library of PostScript procedures designed for the purpose. The very last chapter, which leads to some of the most difficult programming in the book, is concerned with how to triangulate an arbitrary, simple, closed 2D polygon but is motivated by the problem of how to draw figures on 3D surfaces. Certain technical matters are dealt with in several appendixes. I have have also included in an epilogue a few brief remarks on the vast topic of style in mathematical graphics design.

WEB SITE

The book refers to a great deal of PostScript code – samples as well as libraries that the reader may download and incorporate in his or her own programs. Links to this code are to be found at

    http://www.math.ubc.ca/~cass/graphics/manual/

along with other relevant material, including, eventually, software tools for producing PostScript figures more easily. Errors in the text will be flagged and corrected at this site, too. Readers finding errors, or even just with suggestions for improvement, should send e-mail to cass@math.ubc.ca.

ACKNOWLEDGMENTS

These notes have been written over a long period. Right from the start they were made available on the Internet in various intermediate stages of development, and they have benefited greatly from feedback brought about by this availability. I wish to thank above all my patient undergraduate students during those years for their help in improving them. I wish to thank also the many readers who have written to me from all over with encouragement, advice, and error notices. Special thanks are extended to the referees selected by Cambridge University Press as well as its staff for invaluable suggestions. But above all I wish to thank Donald Knuth – not for any personal contribution but for writing all those wonderful papers and books in which he shows us by exhortation and example that interesting mathematics can be found in nearly every corner of the computer.

CHAPTER 1

Getting started in PostScript

In this book we use a program called **Ghostscript**, as well as one of several programs that in turn rely on Ghostscript running behind the scenes, to serve as our PostScript® interpreter and interface. All the programs we use are available without cost through the Internet. Be careful – the language we are writing our programs in is PostScript, and the program used to interpret them is Ghostscript. See Appendix 2 for how to acquire Ghostscript and set up your programming environment.

The interpreter Ghostscript has by itself a relatively primitive user interface that will turn out to be too awkward to use for very long, but learning this interface will give you a valuable feel for the way PostScript works. Furthermore, Ghostscript will continue to serve a useful although limited purpose in debugging as well as animations.

We begin in this chapter by showing how Ghostscript works and then later on explain a more convenient way to produce pictures with PostScript.

1.1 SIMPLE DRAWING

Start up Ghostscript. On Unix® networks this is usually done by typing gs in a terminal window, and on other systems it is usually done by clicking on the icon for Ghostscript. (You can also run Ghostscript in a terminal window – even on Windows® systems; see Appendix 2.) What you get while gs is running are two windows – one a kind of terminal window into which you type commands and from which you read plain text output and the other a graphics window in which things are drawn.

The graphics window, which I will often call the **page**, opens up with a **default coordinate system**. The origin of this coordinate system on a page is at the lower left, and the unit of measurement, which is the same in both horizontal and vertical

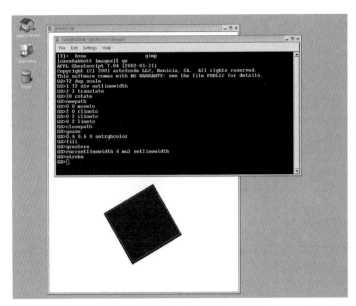

The program Ghostscript running with image and terminal windows showing.

directions, is equal to a **point** exactly $1/72$ of an inch in length. (This **Adobe point** is almost, but not quite, the same as the classical **printer's point**, which measures 72.27 to an inch.) The size of the graphics window will probably be either letter size $(8.5'' \times 11''$ or 612×792 points[2]) or the size of European A4 paper, depending on your locality. As we will see in a moment, the coordinate system can easily be changed so as to arrange x and y units to be anything you want with the origin anywhere in the plane of the page.

When I start up running my local version of Ghostscript in a terminal window I get a display in that window looking like this:

```
AFPL Ghostscript 7.04 (2002-01-31)
Copyright (C) 2001 artofcode LLC, Benicia, CA. All rights reserved.
This software comes with NO WARRANTY: see the file PUBLIC for details.
GS>
```

In short, I am facing the Ghostscript **prompt** GS>, and I am expected to type in commands. Let's start off by drawing a line in the middle of the page. On the left is what the terminal window displays, and on the right is what the graphics window looks like:

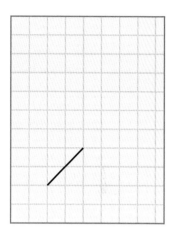

```
GS>newpath
GS>144 144 moveto
GS>288 288 lineto
GS>stroke

GS>
```

(The grid is just there to help you orient yourself and is not displayed in the real window.) The machine produces the prompts here, and everything else is typed by you. The graphics window displays the diagonal line in the figure on the right.

If we look really closely at the line on the screen that comes up, say with a magnifying glass, we'll see a rather jagged image. That's because the screen is made up of pixels with about 75 pixels in an inch. But PostScript is a **scalable** graphics language, which means that if you look at output on a device with a higher resolution than your screen, the effects of pixelization will be seen only at that resolution. Exactly how the computer transforms the directions for drawing a line into a bunch of black pixels is an extremely interesting question but is not one that this book will consider. So, in effect, in this book all lines will be assumed to be ... well, lines – not things that look jagged and ugly – dare I say *pixellated?* – close up.

You draw things in PostScript by constructing **paths**. Any path in PostScript is a sequence of lines and curves. At the beginning, we will work only with lines. In all cases, first you **build** a path and then you actually **draw** it.

- You begin building a path with the command `newpath`. This is like picking up a pen to begin drawing on a piece of paper. In case you have already drawn a path, the command `newpath` also clears away the old path.

- You start the path itself with the command `moveto`. This is like placing your pen at the beginning of your path. In PostScript, things are generally what you might think to be backwards, and so you write down *first* the coordinates of the point to move to and *then* the command.

- You add a line to your path with the command `lineto`. This is like moving your pen on the paper. Again you place the coordinates first and then the command.
- So far you have just built your path. You draw it – that is, make it visible – with the command `stroke`. You have some choice over what color you can draw with, but the color that is used by default is black.

From now on I will usually leave the prompts GS> *out.* Let me repeat what I hope to be clear from this example:

- *PostScript always digests things backwards. The arguments to an operator always go before the operator itself.*

This convention is called **Reverse Polish Notation** (RPN). It will seem somewhat bizarre at first, but you'll get used to it. It is arguable that manual calculations, at least when carried out by those trained in European languages, should have followed RPN conventions instead of the ones used commonly in mathematics. It makes a great deal of sense to apply operations as you write from left to right.

RPN was devised by logicians for purely theoretical reasons, but PostScript is like this for practical reasons of efficiency. There is one great advantage from a user's standpoint: it allows a simple "cut and paste" style of programming.

You would draw a square 2 inches on a side with the command sequence

```
newpath
144 144 moveto
288 144 lineto
288 288 lineto
144 288 lineto
144 144 lineto

stroke
```

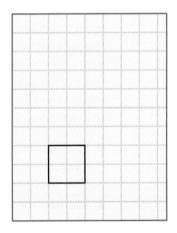

If you type this immediately after the previous command sequence, you will just put the square down on top of the line you have already drawn:

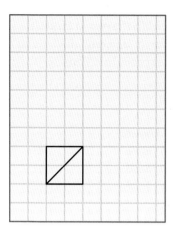

I'll tell you in Section 1.6 how to start over with a clean page. For now, it is important to remember that PostScript *paints over what you have already drawn* just like painting on a canvas. There is no command that erases what has already been drawn.

There are often many different ways to do the same thing in PostScript. Here is a different way to draw the square:

```
newpath
144 144 moveto
144 0 rlineto
0 144 rlineto
-144 0 rlineto
closepath
stroke
```

The commands `rmoveto` and `rlineto` mean motion **relative** to where you were before. The command `closepath` closes up your path back to the last point to which you applied a `moveto`.

A very different effect is obtained with

```
newpath
144 144 moveto
144 0 rlineto
0 144 rlineto
-144 0 rlineto
closepath
fill
```

This just makes a big black square in the same location. *Whenever you build a path, the operations you perform to make it visible are* `stroke` *and* `fill`. The first draws the path; the second fills the region inside it.

You can draw in different shades and colors with two different commands, `setgray` and `setrgbcolor`. Thus,

```
0.5 setgray
newpath
144 144 moveto
144 0 rlineto
0 144 rlineto
-144 0 rlineto
closepath
fill
```

will make a gray square, and

```
1 0 0 setrgbcolor
newpath
144 144 moveto
144 0 rlineto
0 144 rlineto
-144 0 rlineto
closepath
fill
```

will make a red one. The `rgb` here stands for red, green, blue, and for each color you choose a set of three parameters between 0 and 1. Whenever you set a new color, it will generally persist until you change it again. Note that 0 is black and is 1 white. The command x `setgray` is the same as x x x `setrgbcolor`. You can remember that 1 is white by recalling from high school physics that white is made up of all the colors put together.

EXERCISE 1.1. *How would you set the current color to green? Pink? Violet? Orange?*

Filling or stroking a path normally deletes it from the record. So if you want to fill and stroke the same path, you have to be careful. One way of dealing with this is straightforward if tedious – just copy code. If you want to draw a red square with a black outline, you then type

```
1 0 0 setrgbcolor
newpath
144 144 moveto
144 0 rlineto
0 144 rlineto
-144 0 rlineto
closepath
fill
0 setgray
newpath
144 144 moveto
144 0 rlineto
0 144 rlineto
-144 0 rlineto
closepath

stroke
```

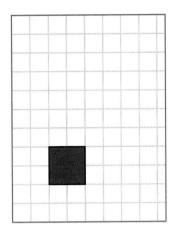

In Section 1.8 we will see a way to produce this figure without redundant typing.

EXERCISE 1.2. *Run Ghostscript. Draw an equilateral triangle near the center of the page instead of a square. Make it 100 points on a side with one side horizontal. First draw it in outline; then fill it in black. Next, make it in turn red, green, and blue with a black outline. (You will have to do a few calculations first. In fact, as we will see in Section 1.4, you can get PostScript to do the calculations.)*

1.2 SIMPLE COORDINATE CHANGES

Working with points as a unit of measure is not for most purposes very convenient. For North Americans, since the default page size is $8.5'' \times 11''$, working with inches usually proves easier. We can change the basic unit of length to an inch by typing

```
72 72 scale
```

which scales up the x and y units by a factor of 72. Scaling affects the current units, and so scaling by 72 is the same as scaling first by 8 and then by 9. This is the way it always works. The general principle here is this:

■ *Coordinate changes are always interpreted relative to the current coordinate system.*

You can scale the x and y axes by different factors, but it is usually a bad idea. Lines are themselves drawable objects of finite width. If scaling is not uniform, the thickness of a line will depend on its direction. Thus, scaling x by 2 and y by 1 has this effect on a square with a thick border:

To be sure to get both scale factors the same, you can also type `72 dup scale`. The command `dup` duplicates the previous entry.

When you scale, you must take into account that the default choice of the width of lines is 1-unit. So if you scale to inches, you will get lines 1-inch wide unless you do something about it. It might be a good idea to add

```
0.01389 setlinewidth
```

when you scale to inches. This sets the width of lines to 1/72 of an inch. A linewidth of 0 is also allowable – it just produces the thinnest possible lines that do not actually vanish. You should realize, however, that on a device of high resolution, such as a 1200 DPI printer, such lines will be nearly invisible. Setting the line width to 0 contradicts the general principle of **device independence** – *you should always aim in PostScript to produce figures that do not in any way depend directly on the particular device on which it will be reproduced.*

EXERCISE 1.3. *How would you scale to centimeters?*

You can also shift the origin.

```
1 2 translate
```

moves the coordinate origin to the right by 1 unit and up by 2 units. The combination

```
72 72 scale
4.25 5.5 translate
```

moves the origin to the center of an 8.5″ × 11″ page.

There is one more simple coordinate change: `rotate`.

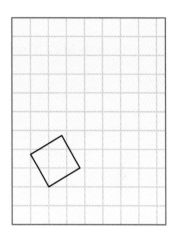

```
144 144 translate
30 rotate
newpath
0 0 moveto
144 0 lineto
144 144 lineto
0 144 lineto
0 0 lineto

stroke
```

The translation is done first because rotation always takes place around the current origin. Note that *PostScript works with angles in degrees.* This will cause us some trouble later on, but for now it is probably A Good Thing.

EXERCISE 1.4. *Europeans use A4 paper. Find out its dimensions and show how to draw a square one centimeter on a side with its center in the middle of an A4 page. (Incidentally, what is the special mathematical property of A4 paper?)*

1.3   COORDINATE FRAMES

It is sometimes not quite so easy to predict the effect of coordinate changes. The secret to doing so is to think in terms of **coordinate frames**. Frames are associated to linear coordinate systems and vice versa. The way to visualize how the coordinate changes `scale`, `translate`, and `rotate` affect drawing is by realizing their effect on the frame of the coordinate system.

A simple frame, with units in centimeters

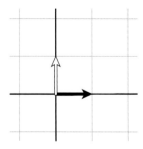

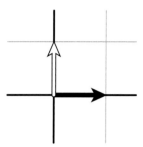

Scaled by $\sqrt{2}$ in both directions

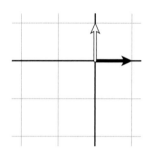

Translated by [1, 1]

Rotated by 45°

There are two fundamental things to keep in mind when wondering how coordinate changes affect drawing.

■ *Coordinate changes affect the current frame in the natural and direct way. That is to say,* 2 2 scale *scales the current frame vectors by a factor of 2, and so on.*

■ *Drawing commands take effect relative to the current frame.*

For example, rotate always rotates the coordinate system around the current origin, which means that it rotates the current coordinate frame. The commands translate, scale, and rotate, when combined in the right fashion, can make any reasonable coordinate change you want (as well as a few you will probably never want). The restriction of "reasonability" here means those that in effect lay down a grid of parallel lines on the plane. As an example, suppose you want to rotate your coordinate system around some point other than the origin. More explicitly, suppose you want to rotate by 45° around the point whose coordinates in the current system are (2, 2). In other words, we want to move the current coordinate frame as at the right.

The way to get this is

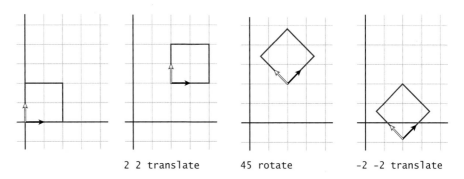

2 2 translate	45 rotate	-2 -2 translate

In other words, to rotate the coordinate system by θ around the point P, you perform in sequence (1) translation by the vector $(P - O)$ from the origin O to P; (2) rotation by θ; (3) translation by $-(P - O)$.

The effect of "zooming in" is rather similar and is analyzed in Appendix 5.

1.4 DOING ARITHMETIC IN PostScript

PostScript is a complete programming language. But with the goal of handling data rapidly, it has only limited built-in arithmetical capabilities. As in many programming languages, both integers and real numbers are of severely limited precision. In some implementations of PostScript, integers must lie in the range $[-32784, 32783]$, and real numbers are only accurate to about seven significant places. This is where the roots of the language in graphics work show up, for normally drawing a picture on a page of reasonable size does not have to be extremely accurate. This limited accuracy is not usually a serious problem, but it does mean you have to be careful.

At any rate, with arithmetical operations as with drawing operations the sequence of commands is backwards from what you might expect. To add two numbers, first enter the numbers followed by the command add. The result of adding numbers is also not quite what you might expect. Here is a sample run in the Ghostscript interpreter:

```
GS>3 4 add
GS<1>
```

What's going on here? What does the <1> mean? Where is the answer?

PostScript uses a **stack** to do its operations. This is an array of arbitrary length that grows and shrinks as a program moves along. The very first item entered is said to be at the **bottom** of the stack, and the last item entered is said to be at its **top**. This is rather like manipulating dishes at a cafeteria. Generally, operations affect only the things towards the top of the stack and compute them without displaying results. For example, the sequence 3 4 add does this:

Entry	What happens	What the stack looks like
3	The number 3 goes onto the stack	3
4	The number 4 goes above the 3 on the stack	3 4
add	The operation add goes above 4 . . .	3 4 add
	then collapses the stack to just a single number	7

You might be able to guess now that the <1> in our run tells us the size of the stack. To display the top of the stack, we type =. If we do this, we get

```
GS>3 4 add
GS<1>=
7
GS>
```

Note that = removes the result when it displays it (as does the similar command ==). An alternative is to type stack or pstack, which displays the entire stack and does not destroy anything on it.

```
GS>3 4 add
GS<1>stack
7
GS<1>
```

The difference between = and == is too technical to explain here, but in practice you should usually use ==. Similarly, you should usually use pstack, which is a bit more capable than stack. There is a third and slightly more sophisticated display operator called print. It differs from the others in that it does not automatically put in a carriage return and can be used to format output. The print command applies basically only to strings, which are put within parentheses. (Refer to Appendix 1 for instructions on how to use print.)

Other arithmetic operations are sub, mul, div. Some of the mathematical functions we can use are sqrt, cos, sin, atan. For example, here is a command sequence computing and displaying $\sqrt{3 * 3 + 4 * 4}$:

```
GS>3 3 mul
GS<1>4 4 mul
GS<2>add
GS<1>sqrt
GS<1>=
5.0
GS>
```

One thing to note here is that the number 5 is written as 5.0, which means that it is a real number, not an integer. PostScript generally treats integers differently from real numbers; only integers can be used as counters, for example. But it can't really tell that the square root of 25 is an integer.

EXERCISE 1.5. *Explain what the stack holds as the calculation proceeds.*

EXERCISE 1.6. *Use Ghostscript to calculate and display $\sqrt{9^2 + 7^2}$.*

Here is a list of nearly all the mathematical operations and functions.

x y add	puts $x + y$ on the stack
x y sub	puts $x - y$ on the stack
x y mul	puts xy on the stack
x y div	puts x/y on the stack
m n idiv	puts the integer quotient of m divided by n on the stack
m n mod	puts remainder after division of m by n on the stack
x neg	puts $-x$ on the stack
y x atan	puts the polar angle of (x, y) on the stack (in degrees)

x sqrt	puts $\sqrt{x}$ on the stack
x sin	puts $\sin x$ on the stack (x in degrees)
x cos	puts $\cos x$ on the stack (x in degrees)
y x exp	puts y^x on the stack
x ln	puts $\ln x$ on the stack
x abs	puts $|x|$ on the stack
x round	puts on the stack the integer nearest x
x floor	puts on the stack the integer just below x
x ceiling	puts on the stack the integer just above x

Note that atan has two arguments and that the order of the arguments is not quite what you might expect. This is commonly done in programming languages.

EXERCISE 1.7. *Do* floor *and* ceiling *return real numbers or integers? What are the floor and ceiling of* 1? −0.5?

EXERCISE 1.8. *Recall that* acos(x) *is the unique angle between* 0° *and* 180° *whose cosine is* x. *Use Ghostscript to find* acos(0.4). *(This will require thinking a bit about the geometry of angles.) (Some versions of PostScript will have* acos *built in. Do not use it, but restrict yourself to getting by with* atan.*)*

In addition to these mathematical functions you will probably find a few elementary commands that move things around on the stack to be useful.

x y exch	exchanges the top two items on the stack to make them y x
x pop	removes the top item on the stack
x dup	makes an extra copy of the top item on the stack

1.5 ERRORS

You will make mistakes from time to time. The default method for handling errors in Ghostscript (and indeed in all PostScript interpreters I am familiar with) is pretty poor. This is not an easy problem to correct, unfortunately. Here is a typical session with a mistake signaled. If you enter

```
GS>5 0 div
```

this is what you will get (more or less) spilled out on the screen:

```
Error: /undefinedresult in --div--
Operand stack:
   5   0
Execution stack:
   %interp_exit   --nostringval--
      --nostringval--    --nostringval--
      %loop_continue   --nostringval--
      --nostringval--    false   --nostringval--
      --nostringval--    --nostringval--
Dictionary stack:
   --dict:592/631--   --dict:0/20--   --dict:34/200--
Current allocation mode is local
GS<2>
```

Holy cow! Your adrenaline level goes way up and your palms break out in sweat. What the $%#?! are you supposed to do now? *Calm down.* The important thing here and with just about all error messages from Ghostscript is that you can ignore all but these first lines:

```
Error: /undefinedresult in --div--
Operand stack:
   5   0
```

which shows you the general category of error and what the stack was like when the error occurred. Here it is division by 0. *It never pays to try too hard to interpret Ghostscript error messages.* The only way to deal with them is to try to figure out where the error occurred and examine your input carefully. There is, alas, essentially just one trick you can use to find out where the error occurred: put lines like (1) = or (location #1) = at various points in your program and try to trace how things go from the way output is displayed. This is simple, but it often helps. The way this works is that (1) denotes the string "1," and = will display it on the terminal. This technique is clumsy, but not much more is possible.

   If you are running Ghostscript, then to recover from an error you probably want to clear the stack completely and start over with the single command clear.

   Incidentally, the way errors are handled by your PostScript interpreter can be modified by suitable embedded PostScript code. In particular, there is a convenient error handler called ehandler.ps available from Adobe via the Internet at `www.adobe.com`. If you have a copy of it in your current directory, you can use it by putting

```
(ehandler.ps) run
```

at the beginning of your program. This will simplify your error messages enormously. You can also arrange for Ghostscript to use it instead of its default error handling, but exactly how depends on which computer you are using. *If you do import* `ehandler.ps` *you must remove all reference to it before sending your work to a printer.*

You should keep in mind that, even for experts, tracking a PostScript program explicitly can be very difficult. One way to write better code in the first place is to include many comments so that someone (usually the programmer!) can tell what the program is doing without following the code itself. In PostScript these are begun with a percent sign %. *All text in a line after a % is ignored.* Thus, the effect of these two lines is the same:

```
(ehandler.ps) run
(ehandler.ps) run        % imports the error handler
```

Another trick for more convenient debugging is to run your program without visual output. On a Unix system this is done with the command `gsnd` (for "**g**host**s**cript **n**o **d**isplay") plus the name of the input file. The messages you get are the same, but this seems to help you concentrate on what Ghostscript is trying to tell you.

1.6   WORKING WITH FILES AND VIEWERS `GhostView` OR `GSView`

Using the Ghostscript interpreter directly shows interesting things, and you should be ready and willing to do it occasionally (for example, when using the `gsnd` option mentioned in the previous section), but it is an extremely inefficient way to produce pictures, mostly because data entered cannot be changed easily, and errors will force you to start all over again. It is much better to work with a Ghostscript **viewer** such as GhostView or GSView, which has a far more convenient interface. Then to produce PostScript programs and visualize them, perform the following sequence of operations:

- *Start up your viewer.*
- *Start up a text editor.*
- *Create or open up in your text editor the file you want to hold your PostScript program. Be sure your file is to be saved as plain text as opposed to one of the special formats many word processors, at least on Windows machines, seem to prefer. This is the*

default with the simple editing program Notepad. Notepad *is capable of handling only short files, however, and sooner or later you will find it inadequate.*

■ *Open up that file from the viewer and see what you've drawn.*

■ *As you make up your program inside the editor, save it from time to time and reopen it in the viewer, where your picture and possibly other messages will be displayed. You can probably set your viewer to reopen the file automatically, whenever it is changed, with a "Watch file" option.*

There are some new features of using files for PostScript programs that you'll have to take into account, but otherwise this works well – indeed, almost painlessly.

■ *At the very beginning of your file you must have the two characters* %!. *This tells your computer that the file is a PostScript file. Sometimes your viewer will be happier if you have a longer line something like* %!PS-Adobe-2.0.

■ *At the end of your file you should have a line with* showpage *on it.*

Neither of these is usually absolutely necessary, but there will be times when both are required. *They will definitely be required if you want to print out your picture on a printer or if you want to import your PostScript file into an image manipulation program in order to turn it into a graphics file of some other format.*

The command showpage displays the current page, at least in some situations, and then starts a new page. Later on you will want to make up files with several pages in them, and each page must have a showpage at the end. There is one tricky feature of showpage, however.

■ *Setting up coordinates, for example scaling, should be done over again on each page.*

There are better and worse ways to deal with this. The best is to put the commands gsave at the beginning of each page and grestore at its end. We will see later exactly what these commands do, but the brief description is that they save and restore the graphics state. Using them as I suggest here just means that every page will start all over in the original graphics environment. Let me repeat this because it is extremely important: *Start each page clean.*

Here, for example, is a complete two-page program:

```
%!

gsave

72 72 scale
1 72 div setlinewidth
4 5 translate
```

```
newpath
0 0 moveto
1 0 lineto
0.5 1 lineto
closepath
stroke

grestore
showpage

gsave

72 72 scale
1 72 div setlinewidth
4 5 translate

newpath
0 0 moveto
1 0 lineto
1 1 lineto
closepath
stroke

grestore
showpage
```

EXERCISE 1.9. *What does this program do?*

By the way, I want to emphasize that spaces, tabs, and line breaks are all the same to PostScript. Thus, in the preceding program I could have written either

```
72 72 scale
1 72 div setlinewidth
4 5 translate
```

or

```
72 72 scale 1 72 div setlinewidth 4 5 translate
```

The only reason to be careful about spaces, tabs, or line breaks in a PostScript program is to make the program readable by humans. This is extremely important to keep in mind even when the only person who reads the program is the one who writes it, for the person who reads a program is *never* the one who writes it. The programmer's brain inevitably changes state in between writing and reading, and it is often *very* difficult for the reader of tomorrow to recall exactly what the writer of today had in mind.

1.7 SOME FINE POINTS

There are several commands that control fine points of the way PostScript draws.

Every line in PostScript has a finite width. Usually this is not apparent, but occasionally it will be. How should the end of a line look (i.e., be capped)? How should two lines join at a corner?

PostScript stores internally a variable `linecap` that controls how lines are capped. This table shows the effect:

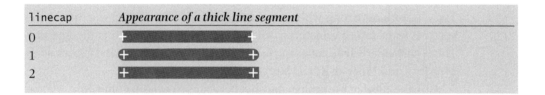

linecap	*Appearance of a thick line segment*
0	
1	
2	

This variable is set with a command sequence like `1 setlinecap`. If it is not explicitly set, it is 0.

How lines join is controlled by an internal variable `linejoin`.

```
9 setlinewidth
% linejoin = 0 by default
newpath
0 0 moveto
0 72 lineto
72 0 lineto

stroke
```

If I add a single line like the following, near the beginning, I get something very slightly different.

```
1 setlinejoin
```

Another possibility is

```
2 setlinejoin
```

There is nothing wrong with any of these pictures, but in some circumstances you will want to use something other than the default, which is the first one. This is particularly true, for example, in drawing three-dimensional figures when linejoin and linecap should both be set to 1.

Also, the effect of closepath may not be what you expect. Compare

```
newpath
0 0 moveto
72 0 rlineto
0 72 rlineto
0 72 lineto
0 0 lineto

stroke
```

and

```
newpath
0 0 moveto
72 0 rlineto
0 72 rlineto
0 72 lineto
closepath

stroke
```

The moral of this is that if you mean to draw a closed path, then use `closepath`. It closes up the path into a seamless path – without break – so that all corners become essentially equivalent. As I will likely often repeat, *programs should reflect concepts* – that is, not depend on accidents to look OK.

EXERCISE 1.10. *Draw in PostScript the following pictures taken, with modifications, from the proof of Proposition I.47 in the standard edition of Euclid's* **Elements**. *Of course you might want to look up Euclid's argument first. One thing to keep in mind is that in drawing a complex figure, each conceptual component in the program should be handled as a unit. A triangle is not just a bunch of line segments but line segments assembled in a particular order and style. You should also think here about about using colors in an intelligent way to help explain Euclid's proof.*

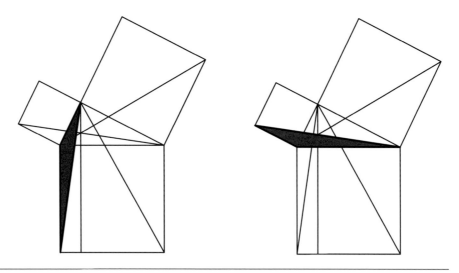

Here and elsewhere, when you are asked to reproduce a picture, you are almost always expected to reproduce its dimensions as closely as you can.

One of the main purposes of this book is to get a reader used to the idea that pictures can and should play a larger role in the exploration and exposition of mathematics. There is nothing like diving in at the deep end – you should take it as dogma here that

■ *The ideal proof has no words and no labels on the figures.*

A good picture – well, a really good one – tells its story without these crutches. The point is that the reader should be able to tell what's going on without having to go back and forth between text and figure and see in a glance what no number of words could express as clearly.

Of course the word "proof" in these circumstances does not have the conventional meaning. Now a proof in mathematics is what it always has been: a rigorous arrangement of statements, each one following logically from previous ones. But in much mathematics, such an arrangement is often flat and without appropriate emphasis. It does not lay out clearly the important points to be dealt with, whereas pictures often do this superbly. The magic of geometry in mathematics, even at the most sophisticated level, is that geometrical concepts are somehow more ... well, more *visible* than others.

EXERCISE 1.11. *Draw a picture of the French flag (blue, white, and red vertical stripes). Do two versions, one with and one without thin black lines separating the different colors and outlining the flag. (I do not know the official aspect ratio, but you should try to find out what it is before you do this exercise.)*

EXERCISE 1.12. *Try a Canadian flag, too. This is somewhat more difficult, but you ought to be able to produce a schematic maple leaf.*

1.8 A TRICK FOR ELIMINATING REDUNDANCY

I mentioned before that if we want to fill a path with one color and then stroke it with another, it is not necessary to build the path a second time. Instead, we can use gsave and grestore in a tricky way. These two operators save and restore the **graphics state**, but the graphics state includes any current paths unfinished as well as the current color. Therefore, we can use

```
newpath
144 144 moveto
144 0 rlineto
0 144 rlineto
-144 0 rlineto
closepath
gsave
1 0 0 setrgbcolor
fill
grestore
0 setgray
stroke
```

to get a red square outlined in black without any redundant code. This trick may seem like much trouble here in this simple case, but when diagrams are more complicated it will prove its worth.

1.9   SUMMARY

The basic commands necessary to draw in PostScript are pretty simple. The commands

```
newpath
moveto
lineto
rmoveto
rlineto
closepath
stroke
fill
```

are used to build and draw paths. The commands

```
translate
scale
rotate
```

allow us to make simple coordinate changes. The commands

```
setlinewidth
setrgbcolor
setgray
setlinejoin
setlinecap
```

allow us to change the attributes of paths to be drawn. Finally,

```
gsave
grestore
showpage
```

lets us put together several pages of drawings.

These are almost all the basic drawing commands. They are fairly simple, and it might be difficult at this stage to understand how one can use them to draw anything complicated. Constructing interesting things to draw can require considerable work. To help us here, we have so far seen only a small set of mathematics functions like cos and the stack manipulation commands dup, exch, and pop. Next we need to learn how to get PostScript to do more sophisticated things for us.

Finally, remember that in PostScript you put data on the stack before you say what to do with those data.

CODE

The file beginning.ps contains several of the figures from this chapter on successive pages.

REFERENCES

1. **PostScript Language – Tutorial and Cookbook**, Adobe Systems, 1985. Available at

   http://www-cdf.fnal.gov/offline/PostScript/BLUEBOOK.PDF

   This is easy and pleasant reading with lots of intriguing examples. Known informally as "the blue book."

2. **PostScript Language – Program Design**, Adobe Systems, 1985. Available at

   http://www-cdf.fnal.gov/offline/PostScript/GREENBK.PDF

   Not quite so useful as the tutorial but still with useful ideas. Known informally as "the green book." Both of these classic manuals are also available from the source at

   http://partners.adobe.com/asn/tech/ps/download/samplecode/ps_psbooks/index.jsp

3. **PostScript Language Reference**, Adobe Systems, Third Edition, 1999. Available at

   http://partners.adobe.com/asn/developer/specifications/postscript.html

   Invaluable. Huge and comprehensive but nonetheless very readable. Known informally as "the red book." Particularly valuable is the alphabetical list of all PostScript operators in addition to a list of operators grouped by function.

4. The file ehandler.ps can be found at

   http://www.adobe.com/support/downloads/detail.jsp?hexID=5396

   and many other places on the Internet.

5. Three good Internet sources for PostScript are

   http://www.prepressure.com/ps/whatis/PSlanguage.htm
   (PostScript humour)

   http://www.vergenet.net/~conrad/fractals/legobrot/
   (PostScript LEGO)

```
http://cgm.cs.mcgill.ca/~luc/PSgeneral.html
```
(Luc Devroye's PostScript pages)

There are many sites on the Internet that display country flags. One apparently good one is

```
http://flagspot.net/flags/
```

These URLs were valid when I wrote this, but might easily change. Use a search engine if so.

## CHAPTER 2

Elementary coordinate geometry

The page on which you draw may, for all practical purposes, be considered a window onto a plane extending uniformly to infinity. We will not look too closely at the assumptions made in this statement but instead rely strongly on intuition depending on visual experience to deduce important facts about this plane.

Using computers to draw requires translating from geometry to numbers – that is, to a coordinate system – and back again. A few basic formulas are used over and over again. It is best to memorize them. Calculating the distance between points whose coordinates are given requires Pythagoras' theorem, which we recall almost at the beginning of this chapter. Before that, however, comes a discussion of the areas of parallelograms; and even before that comes a short note about distinguishing points from vectors. Towards the end of the chapter we consider several results related principally to projections.

For many readers, the results presented in this chapter will be well known. Even for them, however, the use of visual reasoning might be interesting and, in some aspects, novel.

2.1 POINTS AND VECTORS

It is important to distinguish **points** from **vectors** even though a coordinate system assigns a pair of numbers to either a point or a vector. Points are . . . well, points. They possess no attribute other than **position**, and in particular they are (in spite of how they are drawn!) without dimension or size or color or smell or . . . anything other than position. Vectors, on the other hand, have magnitude and direction. They measure **relative position**. It is very important to keep in mind that *both points and vectors are objects independent of which coordinate system is being used*.

Vectors can be added to each other, and they can be multiplied by constants. There is also a kind of limited algebra involving points. If $P = (x_P, y_P)$ and

$Q = (x_Q, y_Q)$ are two points, then there is a unique vector with tail at P and head at Q, whose coordinates are $x_Q - x_P$ and $y_Q - y_P$, describing the relative position of the two points. It is written as $Q - P$. One reason that it is common to confuse points with vectors is that to each point P corresponds the vector $P - O$ from the origin to P. However, if the coordinate system changes, the origin may change. The points themselves won't change, but the vectors they correspond to will likely do so.

If we are given a coordinate system, the vector with coordinates x, y will be $[x, y]$, and the point with those coordinates will be (x, y). The point (x, y) corresponds to the vector $[x, y]$ from the origin to itself, but I repeat that this point and this vector are not the same geometrical object.

If P is a point and v a vector, it makes sense to consider $P + v$ as a point: it is the point Q such that $Q - P = v$. It is the point P **displaced** or **translated** by v. If t is a real number between 0 and 1, then the point t of the way from P to Q is equal to $P + t(Q - P)$ with coordinates $(1 - t)x_P + tx_Q, (1 - t)y_P + ty_Q$. I write it as $(1 - t)P + tQ$. It is a kind of weighted average of P and Q. For example, the point midway between P and Q is $(1/2)P + (1/2)Q$. As we will see Section 6.8, we can also take weighted averages of collections of several points.

In summary, we can subtract two points to get a vector or calculate a weighted average to obtain another point, but the sum of two points or a scalar multiple of a point makes no intrinsic sense.

2.2 AREAS OF PARALLELOGRAMS

Area is a somewhat sophisticated concept not easily analyzed in complete rigor. We are used to thinking of it as a number, but of course the number involved depends on the units involved. It is really a ratio of the area of a region to that of a unit square. So area seems to be a fundamental geometrical characteristic of a region. It is interesting that Euclid starts off Book I of the **Elements** with properties of area that are encapsulated in a few particularly simple axioms. One of these is that *congruent regions have the same area*. Recall that one region is said to be **congruent** to another if it is obtained from it by translation, rotation, or reflection without altering the relative distances between points of the region. Another basic principle is the **additive principle of areas**: *If two regions have the same area and congruent regions are added to each, then the new regions also have the same area.* This leads to the following more general criterion, which is very close to the one used implicitly by Euclid in his treatment of area:

■ (Euclid's first criterion for areas) *Two regions have the same area if they can be chopped into smaller pieces that are congruent.*

This does not allow for a treatment of areas with curved boundaries, but it does allow us to see that

■ *A parallelogram has the same area as a rectangle with the same base and height.*

Why is this true? A proof according to Euclid's criterion must show how to decompose the parallelogram and the rectangle into congruent pieces. In some circumstances, this is simple. The complexity of the decomposition involved depends on how skewed the parallelogram is or how far removed it is from the rectangle it is to be compared with. If it is not too skewed, then we can lop off a triangle at one end of the parallelogram and paste it in at the other to make a rectangle.

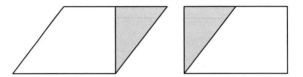

But this means exactly that, in these circumstances, we can decompose the rectangle and the parallelogram into congruent pieces.

If the parallelogram is very skewed, however, then what we lop off at one end is not a triangle, and this argument fails.

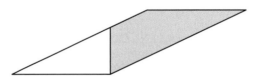

The first, simple argument works when the parallelogram is **mildly skewed** – that is, when the piece chopped off one end is indeed a triangle. This happens when the entire parallelogram fits into the region shown in this figure:

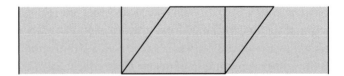

Just about all proofs of the result are the same for mildly skewed parallelograms. There are many different ways to proceed for the rest. Here are a few:

PROOF 2.1. We can get an idea of a possible way to proceed if we again translate the lopped off region to the left and glue it on just as if it were a triangle.

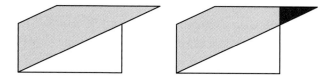

The natural thing to do now is to lop off the bit of triangle at the far right and shift it back again to fill in a rectangle. This finally gives us a way to chop up both the rectangles and the original parallelogram into congruent regions.

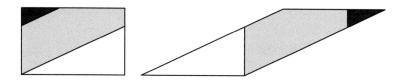

As the parallelogram becomes more and more skewed, the number of pieces the parallelogram gets chopped up into increases, but there is a definite pattern to the way things go. Here are a couple of pictures to show what happens:

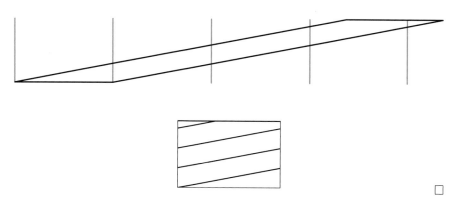

□

EXERCISE 2.1. *Define the* **skew** *of a parallelogram to be the length of the perpendicular projection of its upper left corner onto its base line divided by the length of the base. Count negatively to the left. A parallelogram is a rectangle if and only if its skew is 0. The argument Proof 2.1 shows that if the skew s satisfies $0 < s \leq 1$, then the simple decomposition will prove the claim. Explain by a picture what happens if $-1 \leq s < 0$.*

EXERCISE 2.2. *Explain the argument in the previous exercise by producing figures in PostScript.*

EXERCISE 2.3. *The second group of pictures shows what happens if $1 < s \leq 2$. What about $-2 \leq s < -1$? $2 < s \leq 3$?*

EXERCISE 2.4. *If the skew s satisfies $n < s \leq n + 1$ (n positive), what is the least number of pieces in the decomposition of the parallelogram and rectangle into congruent pieces suggested by the reasoning above?*

EXERCISE 2.5. *The reasoning above has just shown how the decomposition of rectangle and parallelogram works in a few cases, and the preceding exercises have shown how to include a few more cases. Write out in detail a recipe for making congruent decompositions of rectangle and parallelogram that will prove the claim when the skew s satisfies $0 < n < s \leq n + 1$.*

PROOF 2.2. The transformation of a rectangle into a parallelogram with the same base and height is called a **shear**.

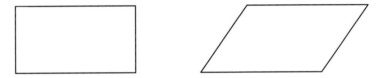

The result we are proving amounts to this:

■ *Shears preserve area.*

A shear can be visualized as a continuous sequence of sliding motions, if you think of the original rectangle as made up of very thin strips piled on top of each other. Like a sliding deck of cards.

In this way, preservation of area under shear becomes intuitive – you can think of the rectangle as an infinite number of horizontal strips piled on top of one another. Shearing it just translates each of these, not changing its area and hence not changing the area of the total figure as it is sheared. This sort of reasoning is not always dependable, but it is valuable nonetheless. Historically, it played an important role in the development of calculus long before the nature of limits was understood clearly. Here, however, it suggests an entirely valid and perhaps the best motivated proof of the result. We don't have to chop up the rectangle into an infinite number of horizontal strips but just enough strips so that each one becomes only a mildly skewed parallelogram when it is sheared.

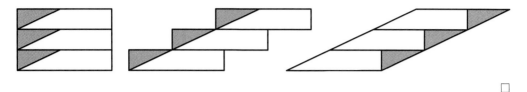

PROOF 2.3. That two parallelograms with the same base and height have the same area is Proposition I.35 in Euclid's **Elements**, but his proof of it depends on the **subtractive principle of areas**: *If congruent regions are taken away from two regions of equal area, then the remaining regions have equal area.*

The simplest of all proofs depends on this principle, but it is not the same as Euclid's. It can be explained in a single pair of diagrams:

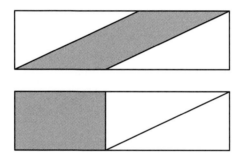

EXERCISE 2.6. *Analyze Euclid's own proof of I.35 by breaking it up into a sequence of pictures.*

EXERCISE 2.7. *Neither the result nor any of these proofs depends on interpreting area as a number nor even on how to compare the area of two distinct rectangles. Make a first step in this direction by explaining in your own words how to construct geometrically, for any rectangle with base b and height a, a square of the same area. (This is II.14 of Euclid's **Elements**).*

2.3   LENGTHS

The principal result concerning lengths is Pythagoras's theorem.

■ *For a right-angled triangle with short sides a and b and long side c, $c^2 = a^2 + b^2$.*

This result, as also the one in the previous section, can be phrased in terms of equality of areas. We erect squares on each of the sides of the given triangle. The theorem asserts neither more nor less than that

■ *The area of the largest square is the sum of the areas of the other two.*

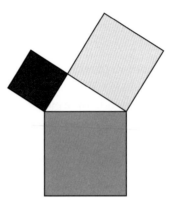

There are many ways to prove Pythagoras's theorem. There is even a book that purports to contain 365 different proofs, one for each day of the year (and includes a few extra). The proof given here is close to Euclid's own (Pythagoras's theorem is I.47 in the **Elements**). I first saw it in a book by Howard Eves, but it probably derives originally from the proof of a generalization of Pythagoras's theorem due to the later Alexandrian mathematician Pappus.

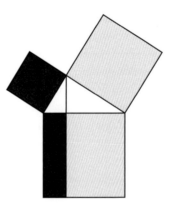

It exhibits a decomposition of the larger square (the "square on the hypotenuse") into rectangles whose areas match the smaller squares (the "squares on the sides"). The proof proceeds by a sequence of shears and translations, which we know to

preserve areas, transforming the rectangles in the large square into the squares on the sides.

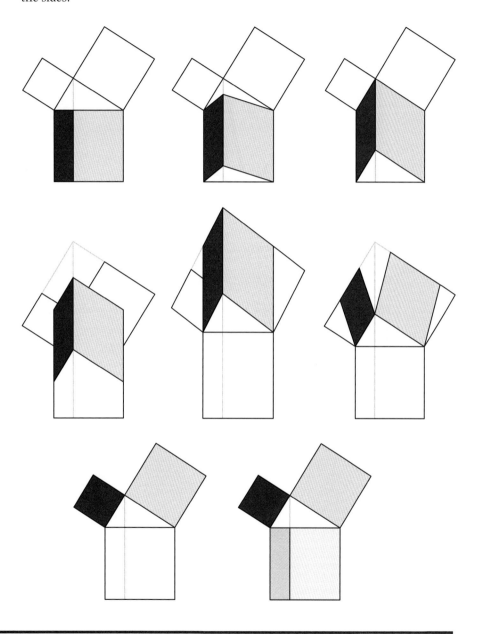

EXERCISE 2.8. *This is very elegant, but if looked at closely there appear to be a few gaps. Find them and fill them in.*

2.4　VECTOR PROJECTIONS

If v is a vector in the plane, then any other vector u can be expressed as the sum of a vector u_0 parallel to v and a vector $u_\perp$ perpendicular to it.

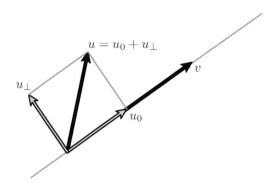

We ask the following question:

- *If $v = [a, b]$ and $u = [x, y]$, how do we calculate u_0?*

The projection u_0 will be a scalar multiple of v, say $u_0 = cv$, and our problem is to calculate c. The length of the projection will be

$$\|u_0\| = |c| \|v\|.$$

So if we know the length $\|u_0\|$, we can calculate $|c| = \|u_0\| / \|v\|$. To obtain the sign of c, we introduce the notion of **signed length**. If the ordinary length of the projection is s, then its signed length (relative to the vector v) is just s if the projection is in the same direction as v, but it is $-s$ if in the opposite direction.

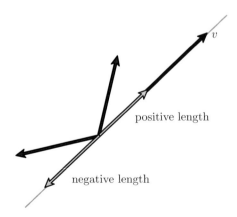

positive length

negative length

We now need to find a formula for the signed length of u_0.

The first observation is that the parallel projection is an additive function of u, which means that if $u = u_1 + u_2$, then the projection of u is the sum of the projections of u_1 and u_2.

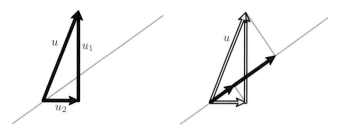

Because u is equal to the sum of its projections onto the x and y axes, it is only necessary to find the signed lengths of the projections of $[x, 0]$ and $[0, y]$ and add them together.

Let's look at the projection of $[x, 0]$.

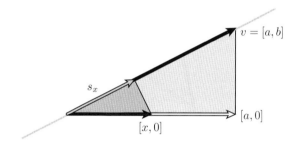

Let s_x be the signed length of the projection of $[x, 0]$ and let $v = [a, b]$. The two triangles in the figure are similar, and so we see that

$$\frac{s_x}{x} = \frac{a}{\|v\|}, \quad s_x = \frac{ax}{\|v\|}.$$

Similarly, if s_y is that of $[0, y]$, then

$$s_y = \frac{by}{\|v\|}.$$

The figure deals with positive lengths, but the final result remains valid for negative ones as well. Hence, the signed length of $u = [x, y]$ is

$$s = s_x + s_y = \frac{ax + by}{\|v\|}.$$

■ *If $u = [x, y]$ and $v = [a, b]$ are vectors, the projection of u onto a line parallel to v is*

$$\frac{ax + by}{\sqrt{a^2 + b^2}} \frac{[a, b]}{\sqrt{a^2 + b^2}} = \left(\frac{ax + by}{a^2 + b^2}\right) [a, b].$$

There is another formula for the signed length. If θ is the angle between u and v, then

$$s = \|u\| \cos\theta.$$

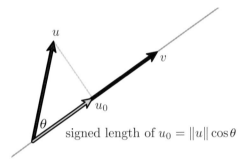

signed length of $u_0 = \|u\| \cos\theta$

If we compare the two formulas, we see that the angle θ between the vectors $v = [a, b]$ and $u = [x, y]$ can be found from this identity:

$$\cos\theta = \frac{ax + by}{\|u\| \|v\|}.$$

If $u = [a, b]$ and $v = [x, y]$, then the numerator of the formula above is called their **dot product**:

$$u \bullet v = ax + by.$$

Thus, the formula for the cosine of the angle between them becomes

$$\cos\theta = \frac{u \bullet v}{\|u\| \|v\|}.$$

The dot product satisfies several simple, formal algebraic rules:

$$cx \bullet y = c(x \bullet y)$$
$$x \bullet cy = c(x \bullet y)$$

$$(x + y) \bullet z = x \bullet z + y \bullet z$$
$$x \bullet x = \|x\|^2,$$

where $\|x\|$ is the length of the vector x, the distance of its head from its tail.

2.5 ROTATIONS

We now look at a new problem:

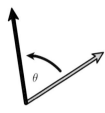

Suppose we start with the vector $v = [x, y]$ and rotate it around the origin by angle θ. What are the coordinates of the new vector?

Rotation by θ

The answer to this depends directly on the answer to the simplest case – that where the angle is $90°$.

■ *If $v = [x, y]$, then rotating v counterclockwise by $90°$ gives us $v_\perp = [-y, x]$.*

This can be seen easily in this picture:

Rotation by $90°$

But now we are in good shape since, from this figure we can deduce the following:

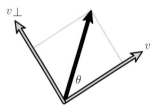

v rotated by θ is $(\cos \theta)v + (\sin \theta)\, v^\perp$

■ *If $v = [x, y]$, then rotating v by θ gives us*

$$(\cos \theta)\, v + (\sin \theta)\, v_\perp = [x \cos \theta - y \sin \theta, \; x \sin \theta + y \cos \theta].$$

This expression can be calculated by a matrix. Rotation by θ takes the vector $[x\ \ y]$ to

$$[x\ \ y]\begin{bmatrix} \cos\theta & \sin\theta \\ -\sin\theta & \cos\theta \end{bmatrix}.$$

In this book, vectors will usually be row vectors, and matrices will multiply them on the right. This is a common convention in computer graphics, as opposed to that in mathematics, and makes especially good sense in dealing with PostScript calculations.

If v is the unit vector $[\cos\alpha,\ \sin\alpha]$ and V_* is v rotated by β, we obtain on the one hand the vector corresponding to angle $\alpha+\beta$ and, on the other, according to the formula for rotations I have just derived, the vector

$$[\cos(\alpha+\beta),\ \sin(\alpha+\beta)] = [\cos\alpha,\ \sin\alpha]\begin{bmatrix} \cos\beta & \sin\beta \\ -\sin\beta & \cos\beta \end{bmatrix}.$$

This gives us the *cosine and sine sum rules*:

$$\cos(\alpha+\beta) = \cos\alpha\cos\beta - \sin\alpha\sin\beta$$
$$\sin(\alpha+\beta) = \sin\alpha\cos\beta + \cos\alpha\sin\beta.$$

2.6   THE COSINE RULE

The **cosine rule** is a generalization of Pythagoras's theorem applying to triangles that are not necessarily right angled.

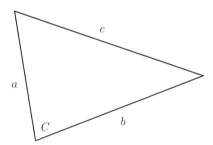

■ (Cosine rule) *In a triangle with sides a, b, c and angle C opposite c*

$$c^2 = a^2 + b^2 - 2ab\cos C.$$

I sketch three proofs. The first is a mixture of algebra and geometry, the second almost purely geometric, and the third almost entirely algebraic.

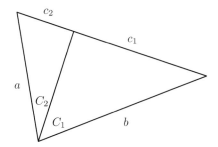

The first one applies the cosine sum formula and Pythagoras's theorem to the diagram above.

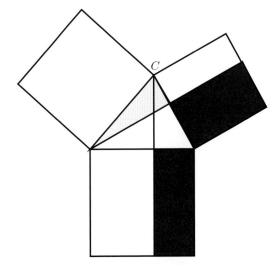

The second generalizes Euclid's proof of Pythagoras's theorem. It begins by showing that the two rectangular areas in the diagram above have equal areas and then finally applies the definition of cosine.

As for the third, it uses the dot product in 2D. Recall that

■ *If the angle between two vectors u and v is θ, then*

$$\cos\theta = \frac{u \bullet v}{\|u\| \|v\|}.$$

Now start by writing the cosine rule in terms of vectors. We want to show that

$$\cos\theta = \frac{\|u - v\|^2 - \|u\|^2 - \|v\|^2}{2\|u\| \|v\|},$$

where u and v are vectors along the sides of the triangle with lengths a and b, and therefore the third side of the triangle is $u - v$. Following the equation above, we reduce

$$\frac{(u-v)\bullet(u-v) - u\bullet u - v\bullet v}{2\|u\|\,\|v\|} = \frac{u\bullet u - 2\,u\bullet v + v\bullet v - u\bullet u - v\bullet v}{2\|u\|\,\|v\|}$$

$$= \frac{u\bullet v}{\|u\|\,\|v\|}$$

$$= \cos\theta \quad \text{sure enough.}$$

EXERCISE 2.9. *Finish the first two proofs. For the second, add several diagrams illustrating the argument.*

2.7 DOT PRODUCTS IN HIGHER DIMENSIONS

The dot product of two vectors in any number n of dimensions greater than two or three is *by definition* the sum of the products of their coordinates:

$$(x_1, x_2, \ldots, x_n)\bullet(y_1, y_2, \ldots, y_n) = x_1 y_1 + x_2 y_2 + \cdots + x_n y_n.$$

The formal rules we have seen to be true in two dimensions hold also in three dimensions and more.

■ *For vectors u and v in two or three dimensions*

$$u\bullet v = \|u\|\,\|v\|\,\cos\theta,$$

where θ is the angle between them.

The proof of this is just the reverse of the third argument in the last section.
    In particular,

■ *The dot product of two vectors is 0 precisely when they are perpendicular to each other.*

Of course it is only in two or three dimensions that we have a geometric definition of the angle between two vectors. In higher dimensions, this formula is used to define that angle algebraically.

2.8 LINES

One way of representing lines in the plane is by means of an equation

$$y = mx + b.$$

But this cannot represent vertical lines, those parallel to the y-axis, which have an equation $x = a$. The uniform way to represent all lines is by means of an equation

$$Ax + By + C = 0.$$

For example, $y - mx - b = 0$ or $x - a = 0$. Lines that are not vertical are those with $B \neq 0$, in which case we can solve for y. The problem with this scheme is that, if $Ax + By + C = 0$ is the equation of a line and c is a nonzero constant, then $cAx + cBy + cC = (cA)x + (cB)y + (cC) = 0$ is also the equation of the same line. This means that the coordinates $[A, B, C]$ of a line are **homogeneous** – only determined up to multiplication by a non-zero scalar. This is the first place in which homogeneous coordinates occur in this book. They will play an extremely important role later on, especially when we come to 3D graphics, and also in understanding how PostScript handles coordinates in 2D.

In the equation $y = mx + b$, both m and b have a geometrical interpretation, that is, m is the slope and b the y-intercept. What is the geometrical significance of A, B, and C in the equation $Ax + By + C = 0$?

Suppose $C = 0$. The equation is $Ax + By = 0$, which can be rewritten

$$[A, B] \bullet [x, y] = 0 .$$

But this is the condition that $[x, y]$ be perpendicular to $[A, B]$. In other words, $[A, B]$ is the direction perpendicular to the line $Ax + By = 0$. In other words, the line $Ax + By = 0$ is the unique line that is (1) perpendicular to the vector $[A, B]$ and (2) passing through the origin.

Now look at the general case $Ax + By + C = 0$. If $P = (x_P, y_P)$ and $Q = (x_Q, y_Q)$ are two points on this line, then

$$Ax_P + By_P + C = 0$$
$$Ax_Q + By_Q + C = 0$$
$$A(x_Q - x_P) + B(y_Q - y_P) = 0,$$

which says that the vector $Q - P$ is perpendicular to $[A, B]$. In other words, we have the following picture:

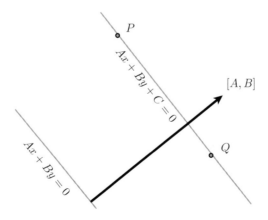

■ *The vector $[A, B]$ is perpendicular to the line $Ax + By + C = 0$.*

What is the meaning of C? If $C = 0$, the line passes through the origin. This makes the following plausible:

■ *The quantity*

$$\frac{-C}{\sqrt{A^2 + B^2}}$$

is the signed distance of the line $Ax + By + C = 0$ from the line $Ax + By = 0$.

EXERCISE 2.10. *Explain why this is true. (Hint: Use projections.)*

EXERCISE 2.11. *Given a line $Ax + By + C = 0$ and a point P, find a formula for the perpendicular projection of P onto the line.*

EXERCISE 2.12. *Given two lines $A_1 x + B_1 y + C_1 = 0$ and $A_2 x + B_2 y + C_2 = 0$, find a formula for the point of intersection.*

EXERCISE 2.13. *Given two points P and Q, find a formula for the line containing them.*

EXERCISE 2.14. *Given a line $Ax + By + C = 0$, find a formula for the line obtained by rotating it by $90°$ around the origin.*

The line $Ax + By + C = 0$ separates the plane into two halves, one in which $Ax + By + C > 0$ and the other in which it is negative. Which side is which?

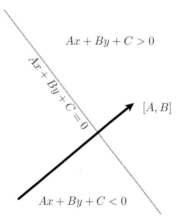

As we cross the line in the direction of $[A, B]$, *the values of* $Ax + By + C$ *change from negative to positive.* This is easy to see indirectly. If $(x, y) = (tA, tB)$, then $Ax + By + C = t(A^2 + B^2) + C$ and for $t \gg 0$ will definitely be positive.

The function $f(x, y) = Ax + By + C$ is called an **affine function**. One useful property of affine functions is this:

- *If P and Q are two points in the plane, t a real number, and f an affine function, then*

$$f\big((1 - t)P + tQ\big) = (1 - t)f(P) + tf(Q).$$

Recall that $(1 - t)P + tQ$ is the weighted average of P and Q. As t varies over all real numbers, this expression produces all points on the line through P and Q. Proving this property is an easy calculation.

EXERCISE 2.15. *An affine function $f(x) = Ax + By + C$ is equal to -4 at $(0, 0)$ and 7 at $(1, 2)$. Where on the line segment between these two points is $f(x) = 0$?*

2.9 CODE

The file `eves-animation.eps` is a page-turning animation of Eves' proof of Pythagoras's theorem.

REFERENCES

1. Euclid, **The Elements**, translated by T. L. Heath. This is available in a commonly found Dover reprint. The part due to Euclid, but not Heath's very valuable comments, is also available on the Internet at

   `http://aleph0.clarku.edu/~djoyce/java/elements/elements.html`

   The commentary used to be found at

   `http://www.perseus.tufts.edu/`

   but at the moment I write this (December 2003) all links to Heath's comments, other than the initial chapters, are unfortunately dead.

2. Elisha S. Loomis, **The Pythagorean Proposition: Its Demonstrations Analyzed and Classified, and Bibliography of Sources for Data of the Four Kinds of Proofs**, National Council of Teachers, Washington, DC, 1968. This book is not wildly exciting and does not offer as much variety one might hope for either, but it is still curious.

3. H. Eves, **In Mathematical Circles**, Prindle, Weber, and Schmidt, 1969. The idea for the sequence of pictures for Pythagoras's theorem is taken from page 75.

CHAPTER 3

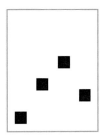

Variables and procedures

At this point, I have explained how to draw only the simplest figures. In particular, I have given no hint of how to use the real programming capability of PostScript.

Before beginning to look at more complex features of the language, place this principle firmly in your mind:

■ *To get good results from PostScript, first get a simple picture up on the screen that comes somewhere close to what you want and then refine it and add to it until it is exactly what you want.*

It is the secret to efficient PostScript programming, because *once you have a picture – any picture – you can often visualize your errors.* Another suggestion is that, since debugging large chunks of PostScript all at once is extremely painful, you want to keep the scope of your errors small. Yet another thing to keep in mind as you develop programs is flexibility. Ask yourself frequently if you might reuse in another drawing what you are doing in this one. We will see how to take advantage of reusable code in an efficient way.

The basic technique of this chapter will be to see how one PostScript program evolves according to this process. Technically, the main ingredients we are going to add to our tool kit are **variables** and **procedures**.

3.1 VARIABLES IN PostScript

The following program draws a square one inch on a side roughly in the middle of a page.

```
%!

72 72 scale
4.25 5.5 translate
1 72 div setlinewidth

newpath
0 0 moveto
1 0 rlineto
0 1 rlineto
-1 0 rlineto
closepath
stroke

showpage
```

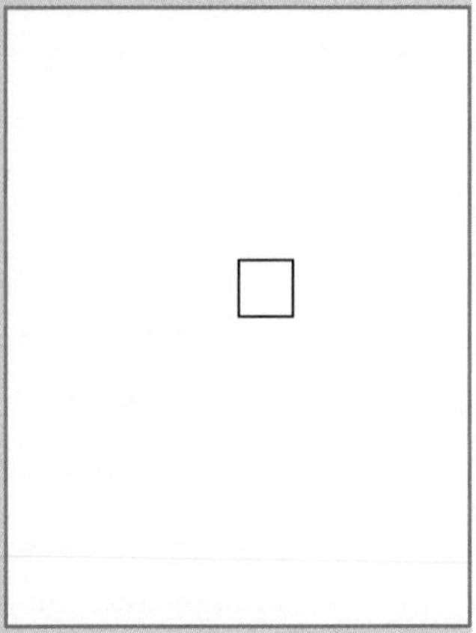

It is extremely simple and frankly not very interesting.

Among other things, it is not very flexible. Suppose you wanted to change the size of the square. You would have to replace each occurrence of "1" with the new size. This is awkward – you might miss an occurrence, at least if your program were more complicated. It would be better to introduce a **variable** s to control the length of the side of the square.

Variables in PostScript can be just about any sequence of letters and symbols. They are **defined** and **assigned values** in statements like this

```
/s 1 def
```

which sets the variable s to be 1. The /s here is the **name** of the variable s. We can't write s 1 def because then the **value** of s would go on the stack and its name would be lost track of, whereas what we want to do is associate the new value 1 with the letter s.

■ *After a variable is defined in your program, any occurrence of that variable will be replaced by the last value assigned to it.*

We will see in Section 3.6 that this is not quite true in certain local environments.

If you attempt to use the name of a variable that has not been defined, you will get an error message about /undefined in ...

If a variable for the side of the square were used, the new program would look like this (I include only the interesting parts from now on):

```
/s 1 def

newpath
0 0 moveto
s 0 rlineto
0 s rlineto
s neg 0 rlineto
closepath
stroke
```

I recall that the command neg replaces anything on the top of the stack by its negative. This code is indeed a bit more flexible than the original because, if you wanted to draw a square of different size, you would have to change only the first line.

Technical remark. Definition and assignment in PostScript look the same and differ only in technical ways. To understand how this works, it is helpful to know how PostScript keeps track of the values of variables. It stores them in a **dictionary**, which is a collection of names and the values assigned to them. There may be several dictionaries currently in use at any given point in a program; they are kept in the **dictionary stack**. When a variable is defined, its name and value are registered in the top dictionary, and this replaces any value it had before. When the variable is encountered in a program, all the dictionaries in use starting at the top of the dictionary stack are searched until the variable's value is found.

3.2 PROCEDURES IN PostScript

Suppose you wanted to draw two squares, one of them at (0, 0) and the other at (0, −1) (that is, just below the first). This can be accomplished straightforwardly through the code:

```
%!

72 72 scale
4.25 5.5 translate
1 72 div setlinewidth
```

```
newpath
0 0 moveto
s 0 rlineto
0 s rlineto
s neg 0 rlineto
closepath
stroke


0 -1 translate


newpath
0 0 moveto
s 0 rlineto
0 s rlineto
s neg 0 rlineto
closepath
stroke


showpage
```

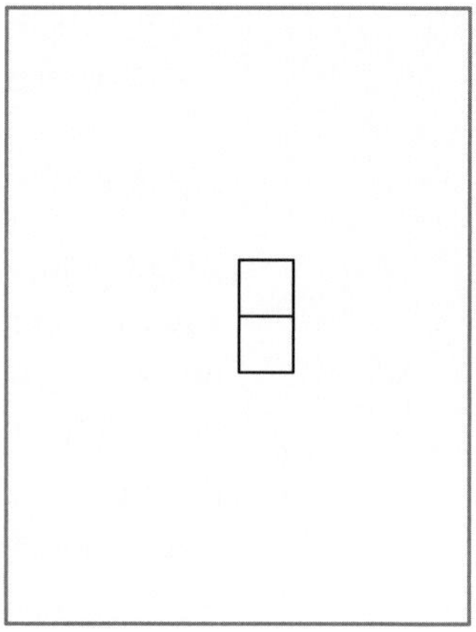

The program, of course, just repeats the part of the program that actually draws the square. Recall that translate shifts the origin of the user's coordinate system in the current units.

Repeating the code to draw the two squares is somewhat inefficient. This technique will lead to much text pasting and turns out to be very prone to error. It is both more efficient and safer to use a PostScript **procedure** to repeat the code for you. A procedure in PostScript is an extremely simple thing.

■ *A procedure in PostScript is just any sequence of commands enclosed in brackets {...}.*

You can assign procedures to variables just as you can assign any other kind of data. When you insert this variable in your program, it is replaced by the sequence of commands inside the brackets. In other words, using a procedure in PostScript to draw squares proceeds in two steps.

(1) Define a procedure, called say draw-square, in the following way:

```
/draw-square {
  newpath
  0 0 moveto
  s 0 rlineto
  0 s rlineto
```

```
  s neg 0 rlineto
  closepath
  stroke
} def
```

At any point in a program after this definition, whenever the expression draw-square occurs, PostScript will simply substitute the lines in between the curly brackets { and }. The effect of calling a procedure in PostScript is always this sort of straightforward substitution.

(2) Call the procedure when it is needed. In this case, the new commands on the page will include the preceding definition and also this:

```
draw-square
0 -1 translate
draw-square
```

Of course if we have done things correctly, the page looks the same as before. But we can now change it easily by mixing several translations and calls to draw-square like this:

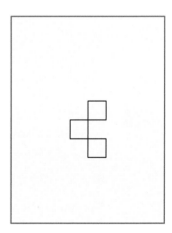

```
draw-square
-1 -1 translate
draw-square
1 -1 translate
draw-square
```

3.3 **KEEPING TRACK OF WHERE YOU ARE**

In the lines of PostScript at the conclusion of the last section, you can easily forget exactly where you are with all those translations. What you might do is translate back again after each translation and drawing operation to restore the original coordinates. But this would become complicated later on in your work, when you will perform several changes of coordinates, and it will be difficult to figure out how to invert them. Instead, you can get PostScript to do the work of remembering where you are. It has a pair of commands that help you do the job easily: gsave

saves the current coordinate system somewhere (together with a few other things like the current line width), and `grestore` brings back the coordinate system you saved with your last `gsave`.

■ *The commands* `gsave` *and* `grestore` *must be used in pairs!*

In this scheme we could write

```
draw-square

gsave
-1 -1 translate
draw-square
grestore

1 -1 translate
draw-square
```

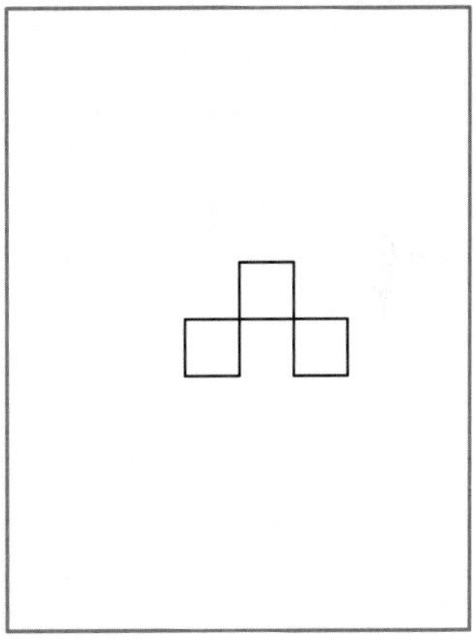

and get something quite different.

To be somewhat more precise, `gsave` saves the **current graphics state** and `grestore` brings it back. The graphics state holds data about coordinates, linewidths, the way lines are joined together, the current color, and more – in effect everything that you can change easily to affect how things are drawn. You might recall that we saw `gsave` and `grestore` earlier, where we used them to set up successive pages correctly, enclosing each page in a pair of `gsave` and `grestore`.

Incidentally, it is usually – but not always – a bad (very bad) idea to change anything in the graphics state in the middle of drawing a path. Effects of this bad practice are often unintuitive and therefore unexpected. There are definite exceptions to this rule, but one must be careful. The problem is to know what parts of the graphic state take effect in various commands. The principal exceptions use `translate` and `rotate` to build paths conveniently. For example, the following sequence builds a square:

```
1 0 moveto 90 rotate
1 0 lineto 90 rotate
1 0 lineto 90 rotate
1 0 lineto 90 rotate
1 0 lineto
```

The commands `rotate` and others change the coordinate system in a figure.

■ *The drawing commands* `lineto` *and others use the coordinate system current when they are applied to build a path in physical coordinates.*

3.4 **PASSING ARGUMENTS TO PROCEDURES**

The definition of the procedure `draw-square` has a variable *s* in it. The variable *s* is not defined in the procedure itself but must be defined before the procedure is used. This is awkward; if you want to draw squares of different sizes, you have to redefine *s* each time you want to use a new size.

For example, if we want two squares of different sizes, we write the code on the left below:

```
/s 1 def
draw-square
/s 2 def

draw-square
```

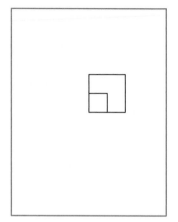

Let me repeat here: *If you want to assign a new value to a variable, you have to define it over again using the* **name** *of the variable, which begins with /*.

Now for a new idea. It is awkward to have to assign a value to *s* every time we want to draw a square. It would be much better if we could just type something like 2 `draw-square` to do the job. This in fact is possible by doing some stack manipulation inside the procedure itself. Let's see – we want to type 2 `draw-square` and draw a square of side 2. This means that the procedure should have access to the item that's put on the stack just before it's called and assign its value to a variable. This requires a trick. Normally we assign a value to a variable by putting the name of the variable on the stack, then the new value, and then calling `def` like this: `/s 2 def`. To assign a value to the variable *s* inside the procedure, we must somehow get the name `/s` on the stack *below* the value on the stack when the procedure is called. The way to do this is to put `/s` on the stack *after* the new value and then call `exch`. The command `exch` exchanges the top two items on the stack. Therefore, 2 `/s exch` makes the stack `/s 2`, and then 2 `/s exch def` has exactly the same effect as /s 2 def. Thus, the lines

```
/draw-square {
  /s exch def
  newpath
  0 0 moveto
  s 0 rlineto
  0 s rlineto
  s neg 0 rlineto
  closepath
  stroke
} def
```

```
2 draw-square
```

do exactly what we want – the side of the square is picked up off the stack, assigned to the variable s, and then used for drawing. The important point is that the procedure itself now handles the assignment of a value to s, and all we do is pass the value of s to the procedure as an **argument** to it by putting it on the stack before the procedure is called. If you are familiar with how programming language compilers work, you will recognize this as what programming languages do to pass arguments – but behind the scenes. The difference is that PostScript does it in the open and effectively forces you to do a bit more work yourself.

If you wanted to draw rectangles with different width and height, you would pass two arguments in a similar way with the code

```
/draw-rectangle {
  /h exch def
  /w exch def
  newpath
  0 0 moveto
  w 0 rlineto
  0 h rlineto
  w neg 0 rlineto
  closepath
  stroke
} def
```

```
2 3 draw-rectangle
```

which draws a rectangle of width 2 and height 3. Notice that the stuff on the stack is removed in the order *opposite* to that in which you placed it there.

■ *In PostScript, the* **arguments** *of a procedure are data that go on the stack before the procedure is called.*

By the way, it seems to me that one of PostScript's principal mistakes in design is the order of the arguments to def. There would be several advantages to having the command work like 2 /s def rather than the way it does. For one thing, reading the arguments of procedures would be more sensible. For another, programmers would be encouraged to make programs more readable. Very often you want to make a very long calculation and then assign the result to a variable. The most common way to do this is to make the calculation and then pick the result off the stack as just described with ... exch def. So here, too, programs would benefit from the change. And they would be more readable because the name of the variable would be close to where it is used. The program would be more *local* in a sense. Locality is one important feature of happy programming.

3.5   PROCEDURES AS FUNCTIONS

A procedure can be a function. That is, it can accept some kind of input, calculate something depending on the input, and pass back or **return** the results of the calculation. There are of course several built-in functions in PostScript such as the mathematical functions neg, add and others, The way these work is that you put **arguments** on the stack, call the function, and then get the **return value** (or values) on the stack. For example, the sequence 30 sin in your program, when the program is executed, puts 30 on the stack (the argument), calls the sin procedure, and leaves 0.5 on the stack (the return value). Some others, like atan, have two arguments. I repeat:

■ *The* **return value** *of a procedure is what it leaves on the stack.*

It can in fact leave much stuff on the stack and can have several return values.
    There is only a formal difference between functions and procedures.

Example. We will make up a procedure hypotenuse that has two arguments and returns the square root of the sum of their squares. In fact, we will see two versions of this. The first will use variables; the second will do all of its work directly with the stack. Both are used in the same way: 3 4 hypotenuse will leave a 5 on the stack. Here is the first, using variables.

```
/hypotenuse {
  /b exch def
  /a exch def
```

```
        a dup mul b dup mul add sqrt
    } def
```

This is reasonably easy to read and understand. There is a problem with it we will deal with in the next section. The second version is not so readable:

```
/hypotenuse {
    dup mul
    exch
    dup mul
    add sqrt
} def
```

This is more efficient than the first; PostScript is generally very efficient when operating directly with stuff on the stack (as opposed to using variables). Still, the cost in terms of readability here is high enough that my general advice is to imitate the first one of these styles rather than the second. If you do want to be more efficient, it is a good idea to add many comments so as to trace what's on the stack.

```
/hypotenuse {    % a b
    dup mul      % a b*b
    exch         % b*b a
    dup mul      % b*b a*a
    add          % b*b+a*a
    sqrt         % the square root of the sum is left on the stack
} def
```

3.6   LOCAL VARIABLES

There is another problem lurking in our present definition of draw-square – that of **variable name conflicts**. If you have a large program with many different figures being drawn in various orders, you might very well have several places in which w and h are used with different meanings. This can cause considerable trouble. The way around this is a technique in PostScript that I suggest you use without trying to understand too much about it in detail. We want the variables we use in a procedure to be **local** to that procedure so that assignments we make to them inside that procedure won't affect other variables with the same name outside the procedure. To do this we add some lines to the procedure

```
/draw-rectangle { 2 dict begin
    /h exch def
```

```
    /w exch def
    newpath
    0 0 moveto
    w 0 rlineto
    0 h rlineto
    w neg 0 rlineto
    0 h neg rlineto
    closepath
    stroke
end } def


2 3 draw-rectangle
```

There are in fact exactly two new lines – one at either end of the procedure. The line
2 dict begin sets up a local variable mechanism, and end restores the original
environment. The 2 is in the statement because we are defining 2 local variables.

■ *You should set up a local variable mechanism in all procedures in which you make
variable definitions.*

One tricky thing to be aware of is that *all variables defined within this pair will be local
variables, and thus it is impossible to change the value of global variables within them*. It
is not usually a good idea to redefine global variables within a procedure anyway.
It is only slightly too extreme to say that you should *never* assign a value to a global
variable inside a PostScript procedure. Again, begin *and* end *have to come in pairs.*
If they don't, the effect will be that a certain block of space in the computer will fill
up. You might get away with it for a while, but unless you are careful, sooner or
later some awful error is bound to occur. Another good rule to follow is that, if you
change coordinates inside a procedure to build a path, for example, you should
restore the original coordinates before it exits unless the explicit purpose of the
procedure is to change coordinates. The general principle is that

■ *Procedures should have as few side effects as possible, and those side effects should
always be explicit.*

Another thing to keep in mind is that, although local variables are indispensable
in procedures, they can in fact be useful anywhere in a program when you want
to avoid name conflicts. Encapsulating a segment of code with 1 dict begin ...
end is often a great way to handle variables safely.

In the programmer's sense, dictionaries are expensive to set up. They are costly
in time. Sometimes, in critical places, it is definitely worthwhile not to introduce
any variables at all in a procedure and to rely entirely on stack manipulations

alone. Or at least use them sparingly. The first of the two procedures

```
/h1 {
  dup mul exch dup mul add sqrt
} def

/h2 { 2 dict begin
/y exch def
/x exch def
x x mul y y mul add sqrt
end } def
```

takes noticeably less time than the second: $10,000,000$ calls on the first take about 5 seconds on my machine, versus about 35 seconds for the second. Recall that to run PostScript programs with no display, as is often useful, you can use Ghostscript in command-line mode as gsnd.

At any rate, don't worry too much about exactly what you have to do to set up local variables. Just copy the pattern without thinking about it. The 2 in 2 dict begin could have been 3 or 4 or 20. In the earliest version of PostScript, it had to be at least as large as the number of variables about to be defined, but in more recent versions a dictionary will expand to whatever size is needed. Thus, even a 1 would be acceptable. In most of my code I am pretty sloppy about this.

3.7 A FINAL IMPROVEMENT

I want to mention here a subtle but valuable point about procedures. It is only rarely a good idea in PostScript to do any actual drawing inside a procedure. This is part of the general principle that the side effects of procedures ought to be severely restricted. Instead, it is usually a good idea to use procedures to build paths without drawing them. Furthermore, it is *always* a good idea to tell in a comment what you have to do to use a procedure and what its effect is. Thus,

```
% Builds a rectangular path with
% first corner at origin.
% On stack at entry: width height
/rectangle { 2 dict begin
  /h exch def
  /w exch def
  0 0 moveto
  w 0 rlineto
  0 h rlineto
```

```
      w neg 0 rlineto
      0 h neg rlineto
      closepath
  end } def

  newpath
  2 3 rectangle
  stroke
```

is the preferable way to use a procedure to draw rectangles. This way you can `fill` them or `clip` them as well as `stroke` them. (We will meet clipping in Section 4.8.) You can also link paths together to make more complicated paths.

CHAPTER 4

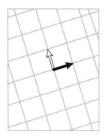

Coordinates and conditionals

We'll take up here a number of drawing problems that require some elementary mathematics and a few new PostScript techniques. These will require that we can interpret absolute location on a page no matter what coordinate changes we have made and therefore motivate a discussion of coordinate systems in PostScript.

At the end we will have, among other things, a complete set of procedures that will draw an arbitrary line specified by its equation. This is not an extremely difficult problem but is one of many whose solution will require understanding how PostScript handles coordinate transformations.

4.1 COORDINATES

The main purpose of PostScript is to draw something – to render it visible by making marks on a physical device of some kind. Every PostScript interpreter is linked to a physical device. Ghostscript running on your computer is linked to your monitor, and printers capable of turning PostScript code into an image possess an interpreter of their own.

When you write a command like `0 0 moveto` or `1 0 lineto` that takes part in constructing a path, the PostScript interpreter immediately translates the coordinates in the command into coordinates more specifically tied to the physical device and then adds these coordinates and the command to a list of commands that will be applied to make marks when the path is finally stroked or filled.

Thus, a PostScript interpreter needs a way to translate the coordinates you write to those required by the physical device; that is, it has to transform the user's coordinates to the ones relevant to the device, and it must internally store some data necessary for this task.

In fact, PostScript deals internally – at least implicitly – with a total of three coordinate systems.

The first is the **physical** coordinate system. This system is the one naturally adapted to the physical device you are working on. Here, even the location of the origin will depend on the device your pictures are being drawn in. For example, on a computer running a version of the Windows operating system, it is apparently always at the lower left. But on a Unix machine, the origin is frequently at the upper left with the y coordinate reading down. The basic units of length in the physical coordinate system are usually the width and the height of one **pixel** (one horizontal, the other vertical), which is the smallest mark that the physical device can deal with. On your computer screen, a pixel is typically $1/75$ of an inch wide and high, whereas on a high-quality laser printer it might be $1/1200$ of an inch on a side. This makes sense because, in the end, every drawing merely colors certain pixels on your screen or printer page.

The second is what I call the **page** coordinate system. This is the one you start up with, in which the origin is at the lower left of the page but the unit of length is 1 Adobe point – equal to $1/72$ of an inch – in each direction. This might be thought of as a kind of ideal physical device.

The third is the system of **user** coordinates. These are the coordinates you are currently using to draw. When PostScript starts up, page coordinates and user coordinates are the same, but certain operations such as `scale`, `translate`, and `rotate` change the relationship between the two. For example, the sequence `72 72 scale` makes the unit in user coordinates equal to an inch. If we then subsequently perform `4.25 5.5 translate`, the translation takes place in the new user coordinates, and thus the origin is shifted up and right by several inches. This is the same as if we had done `306 396 translate` before we scaled to inches, since $306 = 4.25 \cdot 72$ and $396 = 5.5 \cdot 72$.

At all times, PostScript maintains internally the data required to change from user to physical coordinates and implicitly the data required to change from user to page coordinates as well. The formula used to transform coordinates from one system to another involves six numbers and looks like this:

$$x_{\text{physical}} = a\, x_{\text{user}} + c\, y_{\text{user}} + e$$
$$y_{\text{physical}} = b\, x_{\text{user}} + d\, y_{\text{user}} + f$$

PostScript stores these six numbers a, b, and so on in a data structure we will see more of in Section 4.2.

Coordinate changes like this are called **affine coordinate transformations**. An affine transformation is a combination of a **linear transformation** with a shift of the origin. One good way to write the formula for an affine coordinate transformation is in terms of a matrix:

$$[x_\bullet \quad y_\bullet] = [x \quad y] \begin{bmatrix} a & b \\ c & d \end{bmatrix} + [e \quad f].$$

The 2×2 matrix is called the **linear** component of the coordinate transformation, and the vector added on is called its **translation** component. The latter records where the origin is transformed to, and the linear component records how relative positions are transformed.

Affine transformations are characterized by the geometric property that they take lines to lines. They also have the stronger property that they take parallel lines to parallel lines. Linear transformations have in addition the property that they take the origin to itself. The following can indeed be rigorously proven:

■ *An affine transformation of the plane takes lines to lines and parallel lines to parallel lines. Conversely, any transformation of the plane with these properties is an affine transformation.*

Later on, we will see also a third class of transformations of the plane called **projective transformations** (related to perspective viewing). These are not built into PostScript as affine transformations are.

I have said that PostScript uses six numbers to transform user coordinates to physical coordinates and that these six numbers change if you put in commands like `scale`, `translate`, and `rotate`. It might be useful to track how things go as a program proceeds. In this example, I'll assume that the physical coordinates are the same as page coordinates. When PostScript starts up, the user coordinates (x_0, y_0) are the same as page coordinates:

$$x_{\text{page}} = x_0$$

$$y_{\text{page}} = y_0$$

or

$$[x_{\text{page}} \quad y_{\text{page}}] = [x_0 \quad y_0] \begin{bmatrix} 1 & 0 \\ 0 & 1 \end{bmatrix}.$$

If we perform `306 396 translate`, we find ourselves with new coordinates (x_1, y_1). The page coordinates of the new origin are $(306, 396)$. The command sequence `x y moveto` now refers to the point which, in page coordinates, is $(x + 306, y + 396)$. We thus have

$$[x_0 \quad y_0] = [x_1 \quad y_1] + [306 \quad 396]$$

or

$$[x_{\text{page}} \quad y_{\text{page}}] = [x_1 \quad y_1] \begin{bmatrix} 1 & 0 \\ 0 & 1 \end{bmatrix} + [306 \quad 396].$$

If we now perform `72 72 scale`, we find ourselves with a new coordinate system (x_2, y_2). The origin doesn't change, but the command `1 1 moveto` moves to the point that was $(72, 72)$ a moment ago and $(72 + 306, 72 + 396)$ before that.

$$[x_1 \quad y_1] = [x_2 \quad y_2] \begin{bmatrix} 72 & 0 \\ 0 & 72 \end{bmatrix}$$

or

$$[x_{\text{page}} \quad y_{\text{page}}] = [x_2 \quad y_2] \begin{bmatrix} 72 & 0 \\ 0 & 72 \end{bmatrix} + [306 \quad 396].$$

If we now put in 90 `rotate` we have new coordinates (x_3, y_3). The command
1 1 `moveto` moves to the point that was $(-1, 1)$ a moment ago.

$$[x_2 \quad y_2] = [x_3 \quad y_3] \begin{bmatrix} 0 & 1 \\ -1 & 0 \end{bmatrix}$$

or

$$[x_{\text{page}} \quad y_{\text{page}}] = [x_3 \quad y_3] \begin{bmatrix} 0 & 1 \\ -1 & 0 \end{bmatrix} \begin{bmatrix} 72 & 0 \\ 0 & 72 \end{bmatrix} + [306 \quad 396]$$

$$= [x_3 \quad y_3] \begin{bmatrix} 0 & 72 \\ -72 & 0 \end{bmatrix} + [306 \quad 396].$$

The point is that coordinate changes accumulate. We'll see later (in Section 4.5)
more about how this works.

4.2   HOW PostScript STORES COORDINATE TRANSFORMATIONS

The data determining an affine coordinate change

$$[x_\bullet \quad y_\bullet] = [x \quad y] \begin{bmatrix} a & b \\ c & d \end{bmatrix} + [e \quad f]$$

are stored in PostScript in an **array** [a b c d e f] of length six, which it calls a
matrix. (We'll look at arrays in more detail in the next chapter, and we'll see in
a moment why the word "matrix" is used.) PostScript has several operators that
allow you to find out what these arrays are and to manipulate them.

Command sequence	Effect
`matrix currentmatrix`	Puts the current transformation matrix on the stack

There are good reasons why this is a little more complicated than you might expect. The **current transformation matrix** or **CTM** holds data giving the current
transformation from user to physical coordinates. Here the command `matrix` puts

an array [1 0 0 1 0 0] on the stack (the identity transformation), and `current-matrix` stores the current transformation matrix entries in this array. The way this works might seem a bit strange, but it restricts us from manipulating the CTM too carelessly.

For example, we might try this at the beginning of a program and get

```
matrix currentmatrix ==
[1.33333 0 0 1.33333 0 0]
```

The difference between = and ==, which both pop and display the top of the stack, is that the second displays the contents of arrays, which is what we want to do here, whereas = does not.

The output we get here depends strongly on what kind of machine we are working on. The one here was a laptop running Windows 95. Windows 95 puts a coordinate system in every window and positions the origin at lower left with one unit of length equal to the width of a pixel. The origin is thus the same as that of the default PostScript coordinate system, but the unit size might not match. In fact, we can read off from what we see here that on my laptop that one Adobe point is 4/3 pixels wide.

EXERCISE 4.1. *What is the screen resolution of this machine in DPI (dots per inch)?*

As we perform various coordinate changes, the CTM will change drastically. But we can always recover what it was at startup by using the command `defaultmatrix`.

`matrix defaultmatrix`   Puts the original transformation matrix on the stack

The **default matrix** holds the transformation from page to physical coordinates. Thus, at the start of a PostScript program, the commands `defaultmatrix` and `currentmatrix` will have the same effect.

We can solve the equations

$$[x_\bullet \quad y_\bullet] = [x \quad y] \begin{bmatrix} a & b \\ c & d \end{bmatrix} + [e \quad f]$$

for x and y in terms of $x_\bullet$ and $y_\bullet$. The transformation taking $x_\bullet$ and $y_\bullet$ to x and y is the transformation **inverse** to the original. Explicitly, from

$$P_\bullet = PA + v$$

we get

$$P = P_\bullet A^{-1} - v A^{-1},$$

and so it is again an affine transformation. PostScript has an operator that calculates it. The composition of two affine transformations is also an affine transformation that PostScript can calculate:

`M matrix invertmatrix`	Puts the transformation matrix inverse to M on the stack
`A B matrix concatmatrix`	Puts the product AB on the stack

Here, M, A, B are transformation "matrices" – arrays of six numbers.

Thus, the following procedure returns the "matrix" corresponding to the transformation from user to page coordinates:

```
/user-to-page-matrix {
  matrix currentmatrix
  matrix defaultmatrix
  matrix invertmatrix
  matrix concatmatrix
} def
```

To see why, let C be the matrix transforming user coordinates to physical coordinates, which we can read off with the command currentmatrix. Let D be the default matrix we get at start-up. The transformation from the current user coordinate system to the original one is therefore the matrix product $C\,D^{-1}$: C takes user coordinates to physical ones, and the inverse of D takes these back to page coordinates.

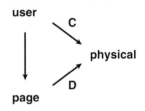

For most purposes, you will not need to use invertmatrix or concatmatrix. The transformation from user to physical coordinates, and back again, can be carried out explicitly in PostScript with the commands transform and itransform. At any point in a program, the sequence x y transform will return the physical coordinates of the point whose user coordinates are (x, y), and the sequence x y itransform will return the user coordinates of the point whose physical coordinates are (x, y). If m is a matrix, then

```
x y m transform
```

will transform (x, y) by m, and similarly for x y m itransform. Thus,

```
x y transform matrix defaultmatrix itransform
```

will return the page coordinates of (x, y).

The operators transform and itransform are somewhat unusual among PostScript operators in that their effect depends on the type of data on the stack when they are used. You, too, can define procedures that behave like this by using the type operator to see what kind of stuff is on the stack before acting.

EXERCISE 4.2. *Write a procedure* page-to-user *with two arguments* x y *that returns on the stack the user coordinates of the point whose page coordinates are* x y. *Also do this for* user-to-page.

4.3   PICTURING THE COORDINATE SYSTEM

In trying to understand how things work with coordinate changes, it might be helpful to show some pictures of the two coordinate systems, the user's and the page's, in different circumstances. (Recall that the page coordinate system is for a kind of imaginary physical device.)

The basic geometric property of an affine transformation is that it takes parallelograms to parallelograms, and so does its inverse. Here are several pictures of how the process works. On the left in each figure is a sequence of commands, and on the right is how the resulting coordinate grid lies over the page.

```
72 72 scale
4.25 5.5 translate
```

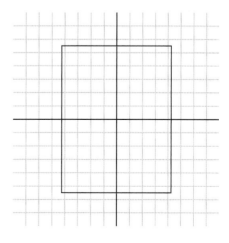

In this figure, the user unit is 1 inch, and a grid at that spacing is drawn at the right.

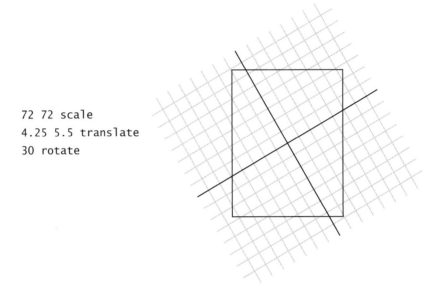

```
72 72 scale
4.25 5.5 translate
30 rotate
```

A line drawn in user coordinates is drawn on the page after rotation of 30° relative to what it was drawn as before.

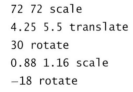

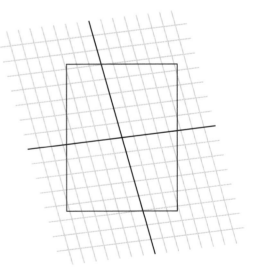

```
72 72 scale
4.25 5.5 translate
30 rotate
0.88 1.16 scale
−18 rotate
```

A combination of rotations and scales can have odd effects after a scale in which the *x*-scale and the *y*-scale are distinct. This is nonintuitive but happens because, after such a scale, rotations take place in that skewed metric.

4.4   MOVING INTO THREE DIMENSIONS

It turns out to be convenient, when working with affine transformations in two dimensions, to relate them to linear transformations in three dimensions.

The basic idea is to associate to each point (x, y) in 2D the point $(x, y, 1)$ in 3D. In other words, we are embedding the two-dimensional (x, y) plane in three dimensions by shifting it up one unit in height. The main point is that the *affine* 2D transformation

$$[x_\bullet \quad y_\bullet] = [x \quad y] \begin{bmatrix} a & b \\ c & d \end{bmatrix} + [e \quad f]$$

can be rewritten in terms of the *linear* 3D transformation

$$[x_\bullet \quad y_\bullet \quad 1] = [x \quad y \quad 1] \begin{bmatrix} a & b & 0 \\ c & d & 0 \\ e & f & 1 \end{bmatrix}.$$

You should check by explicit calculation to see that this is true. In other words, the special 3×3 matrices of the form

$$\begin{bmatrix} a & b & 0 \\ c & d & 0 \\ e & f & 1 \end{bmatrix}$$

are essentially affine transformations in two dimensions if we identify 2D vectors $[x, y]$ with 3D vectors $[x, y, 1]$ obtained by tacking on 1 as last coordinate. This identifies the usual 2D plane, not with the plane $z = 0$ but with $z = 1$. One advantage of this association is that, if we perform two affine transformations successively

$$[x \quad y] \longmapsto [x_1 \quad y_1] = [x \quad y] \begin{bmatrix} a & b \\ c & d \end{bmatrix} + [e \quad f]$$

$$[x_1 \quad y_1] \longmapsto [x_2 \quad y_2] = [x_1 \quad y_1] \begin{bmatrix} a_1 & b_1 \\ c_1 & d_1 \end{bmatrix} + [e_1 \quad f_1],$$

then the composition of the two corresponds to the product of the two associate 3×3 matrices

$$\begin{bmatrix} a & b & 0 \\ c & d & 0 \\ e & f & 1 \end{bmatrix} \begin{bmatrix} a_1 & b_1 & 0 \\ c_1 & d_1 & 0 \\ e_1 & f_1 & 1 \end{bmatrix}.$$

This makes the rule for calculating the composition of affine transformations relatively easy to remember.

There are other advantages to moving 2D points into 3D. A big one involves calculating the effect of coordinate changes on the equations of lines. For example, the line $x + y - 100 = 0$ is visible in the default coordinate system as a line that crosses the screen at lower left from $(0, 100)$ to $(100, 0)$. If we then `100 50 trans-late` we will be operating in a new coordinate system where the new origin is at the point that used to be $(100, 50)$. The line that was formerly $x + y - 100 = 0$ will have a new equation in the new coordinates.

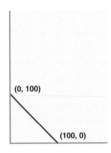
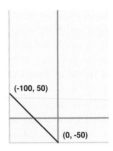

The line itself is something that doesn't change – in practical terms it's that same collection of pixels at the lower left – but its representation by an equation will change. What is the new equation? We have

$$x_\bullet = x - 100, \quad y_\bullet = y - 50,$$

and so

$$x + y - 100 = (x_\bullet + 100) + (y_\bullet + 50) - 100 = x_\bullet + y_\bullet + 50.$$

Now let's look at the general case. The equation of the line

$$Ax + By + C = 0$$

can be expressed purely in terms of matrix multiplication as

$$[x \quad y \quad 1] \begin{bmatrix} A \\ B \\ C \end{bmatrix} = 0.$$

This makes it simple to answer the following question:

■ *Suppose we perform an affine coordinate change*

$$[x \quad y] \longmapsto [x_\bullet \quad y_\bullet] = [x \quad y] \begin{bmatrix} a & b \\ c & d \end{bmatrix} + [e \quad f].$$

If the equation of a line in (x, y) coordinates is $Ax + By + C = 0$, what is it in terms of $(x_\bullet, y_\bullet)$ coordinates?

For example, if we choose new coordinates to be the old ones rotated by $90°$, then the old x-axis becomes the new y-axis, and vice versa.

How do we answer the question? The equation we start with is

$$[x \quad y \quad 1] \begin{bmatrix} A \\ B \\ C \end{bmatrix} = 0.$$

We have

$$[x_\bullet \ y_\bullet \ 1] = [x \ y \ 1] \begin{bmatrix} a & b & 0 \\ c & d & 0 \\ e & f & 1 \end{bmatrix}, \qquad [x \ y \ 1] = [x_\bullet \ y_\bullet \ 1] \begin{bmatrix} a & b & 0 \\ c & d & 0 \\ e & f & 1 \end{bmatrix}^{-1} ;$$

therefore,

$$Ax + By + C = [x \quad y \quad 1] \begin{bmatrix} A \\ B \\ C \end{bmatrix}$$

$$= [x_\bullet \quad y_\bullet \quad 1] \begin{bmatrix} a & b & 0 \\ c & d & 0 \\ e & f & 1 \end{bmatrix}^{-1} \begin{bmatrix} A \\ B \\ C \end{bmatrix}$$

$$= [x_\bullet \quad y_\bullet \quad 1_\bullet] \begin{bmatrix} A_\bullet \\ B_\bullet \\ C_\bullet \end{bmatrix}$$

$$= A_\bullet x_\bullet + B_\bullet y_\bullet + C_\bullet$$

if

$$\begin{bmatrix} A_\bullet \\ B_\bullet \\ C_\bullet \end{bmatrix} = \begin{bmatrix} a & b & 0 \\ c & d & 0 \\ e & f & 1 \end{bmatrix}^{-1} \begin{bmatrix} A \\ B \\ C \end{bmatrix}.$$

In summary,

■ *If we change coordinates according to the formula*

$$[x_\bullet \quad y_\bullet \quad 1] = [x \quad y \quad 1] \begin{bmatrix} a & b & 0 \\ c & d & 0 \\ e & f & 1 \end{bmatrix},$$

then the line $Ax + By + C = 0$ *is the same as the line* $A_\bullet x_\bullet + B_\bullet y_\bullet + C_\bullet = 0$, *where*

$$\begin{bmatrix} A_\bullet \\ B_\bullet \\ C_\bullet \end{bmatrix} = \begin{bmatrix} a & b & 0 \\ c & d & 0 \\ e & f & 1 \end{bmatrix}^{-1} \begin{bmatrix} A \\ B \\ C \end{bmatrix}.$$

To go with this result, it is useful to know that

$$\begin{bmatrix} A & 0 \\ v & 1 \end{bmatrix}^{-1} = \begin{bmatrix} A^{-1} & 0 \\ -v\,A^{-1} & 1 \end{bmatrix},$$

which you can check by multiplying. Here A is a 2×2 matrix and v a row vector. It is also useful to know that

$$\begin{bmatrix} a & c \\ b & d \end{bmatrix}^{-1} = \begin{bmatrix} d/\Delta & -c/\Delta \\ -b/\Delta & a/\Delta \end{bmatrix}, \qquad \Delta = ad - bc.$$

Here is a PostScript procedure that has two arguments, a "matrix" M and an array of three numbers A, B, and C that returns on the stack the array of three numbers $A_\bullet$, $B_\bullet$, $C_\bullet$ that go in the equation for the transform under M of the line $Ax + By + C = 0$. The procedure starts by removing its components. It begins this with a short command sequence aload pop, which spills out the array onto the stack in order. (The operator aload puts the array itself on the stack as well, and pop gets rid of it.)

```
/transform-line { 1 dict begin
  aload pop
  /C exch def
  /B exch def
  /A exch def
  /M exch def
  /Minv M matrix  invertmatrix def
  [
    A Minv 0 get mul B Minv 1 get mul add
    A Minv 2 get mul B Minv 3 get mul add
    A Minv 4 get mul B Minv 5 get mul add C add
  ]
  end } def
```

This is the first time arrays have been dealt with directly in this book. To understand this program, you should know that

(1) the items in a PostScript array are indexed starting with 0;

(2) if A is an array in PostScript, then A i get returns the ith element of A.

EXERCISE 4.3. *If we set*

$$x_\bullet = x + 3, \quad y_\bullet = y - 2,$$

what is the equation in $(x_\bullet, y_\bullet)$ *of the line* $x + y = 1$?

EXERCISE 4.4. *If we set*

$$x_\bullet = -y + 3, \quad y_\bullet = x - 2,$$

what is the equation in $(x_\bullet, y_\bullet)$ *of the line* $x + y = 1$?

EXERCISE 4.5. *If we set*

$$x_\bullet = x - y + 1, \quad y_\bullet = x + y - 1,$$

what is the equation in $(x_\bullet, y_\bullet)$ *of the line* $x + y = 1$?

4.5 HOW COORDINATE CHANGES ARE MADE

Let's look more closely at how PostScript makes coordinate changes.

Suppose we are working with a coordinate system $(x_\bullet, y_\bullet)$. After a scale change

```
2 3 scale
```

we'll have new coordinates (x, y). What is the relationship between new and old? One unit along the x-axis in the new system spans two in the old, and one along the y-axis spans three of the old. In other words

$$x_\bullet = 2x$$
$$y_\bullet = 3y,$$

or

$$[x_\bullet \quad y_\bullet] = [x \quad y] \begin{bmatrix} 2 & 0 \\ 0 & 3 \end{bmatrix},$$

or

$$[x_\bullet \quad y_\bullet \quad 1] = [x \quad y \quad 1] \begin{bmatrix} 2 & 0 & 0 \\ 0 & 3 & 0 \\ 0 & 0 & 1 \end{bmatrix}.$$

If $T_\bullet$ is the original CTM, then

$$[x_\bullet \quad y_\bullet \quad 1] T_\bullet = [x_{\text{physical}} \quad y_{\text{physical}} \quad 1],$$

and for the new coordinates we have

$$[x \quad y \quad 1] \begin{bmatrix} 2 & 0 & 0 \\ 0 & 3 & 0 \\ 0 & 0 & 1 \end{bmatrix} T_\bullet = [x_{\text{physical}} \quad y_{\text{physical}} \quad 1],$$

and thus the new CTM is

$$T = \begin{bmatrix} 2 & 0 & 0 \\ 0 & 3 & 0 \\ 0 & 0 & 1 \end{bmatrix} T_\bullet.$$

This is the general pattern. A 3×3 matrix corresponds to each of the basic coordinate-changing commands in PostScript according to the transformation from new coordinates to old ones:

a b scale $\qquad \begin{bmatrix} a & 0 & 0 \\ 0 & b & 0 \\ 0 & 0 & 1 \end{bmatrix}$

x rotate $\qquad \begin{bmatrix} \cos x & \sin x & 0 \\ -\sin x & \cos x & 0 \\ 0 & 0 & 1 \end{bmatrix}$

a b translate $\qquad \begin{bmatrix} 1 & 0 & 0 \\ 0 & 1 & 0 \\ a & b & 1 \end{bmatrix}.$

The effect of applying one of these commands is to multiply the current transformation matrix on the *left* by the appropriate matrix.

You can perform such a matrix multiplication explicitly in PostScript. The command sequence

```
[a b c d e f] concat
```

has the effect of multiplying the CTM on the left by

$$\begin{bmatrix} a & b & 0 \\ c & d & 0 \\ e & f & 1 \end{bmatrix}.$$

You will rarely want to do this. Normally, a combination of rotations, scales, and translations will suffice. In fact, every affine transformation can be expressed as such a combination, although it is not quite trivial to find it.

EXERCISE 4.6. *After operations*

```
72 72 scale
4 5 translate
30 rotate
```

what is the user-to-page coordinate transformation matrix?

EXERCISE 4.7. *Any 2×2 matrix A may be expressed as*

$$A = R_1 \begin{bmatrix} s & \\ & t \end{bmatrix} R_2,$$

where R_1 and R_2 are rotation matrices. How do we find these factors? Write

$$
\begin{aligned}
{}^t A\, A &= {}^t R_2\, {}^t S\, {}^t R_1\, R_1\, S\, R_2 \\
&= R_2^{-1} S^2 R_2
\end{aligned}
$$

because $R^{-1} = {}^t R$ for a rotation matrix R and ${}^t S = S$. This means that the diagonal entries of S^2 are the eigenvalues of the symmetric matrix ${}^t A\, A$ and that the rows of R_2 are its eigenvectors. To find S from S^2, the signs of square roots must be chosen. Describe how to do this and then how to find R_1. Find this factorization for the shear

$$\begin{bmatrix} 1 & 1 \\ & 1 \end{bmatrix}.$$

Then write a PostScript program that combines rotations and scales to draw a sheared unit circle around the origin. What shape do you get? Why?

4.6   **DRAWING INFINITE LINES: CONDITIONALS IN PostScript**

We have seen that a line can be described by an equation

$$Ax + By + C = 0.$$

Recall that the geometrical meaning of the constants A and B is that the direction $[A, B]$ is perpendicular to the line as long as the coordinate system is an **orthogonal** one, where the x and y units are the same and their axes perpendicular.

The problem we now want to take up is this:

■ *We want to make up a procedure with a single argument, an array of three coordinates* [A B C] *whose effect is to draw the part of the line $Ax + By + C = 0$ visible on the page.*

I recall that an **argument** for a PostScript procedure is an item put onto the stack just before the procedure itself is called. I recall also that generally the best way to use procedures in PostScript to make figures is to use them to build paths, not to do any of the actual drawing. Thus, the procedure we are to design, which I will call mkline, will be used like this

```
newpath
[1 1 1] mkline
stroke
```

if we want to draw the visible part of the line $x + y + 1 = 0$.

One reason this is not quite a trivial problem is that we are certainly not able to draw the entire infinite line. There is essentially only one way to draw parts of a line in PostScript and that is to use moveto and lineto to draw a segment of the line given two points on it. Therefore, the mathematical problem we are looking at is this: *If we are given A, B, and C, how can we find two points P and Q with the property that the line segment between them contains all the visible part of the line $Ax + By + C = 0$?* We do not have to worry about whether or not the segment PQ coincides exactly with the visible part; PostScript will handle naturally the problem of ignoring the parts that are not visible. Of course the visible part of the line will exit the page usually at two points, and if we want to do a really professional job, we can at least think about the more refined problem of finding them. But we'll postpone this approach for now.

Here is the rough idea of our approach: (1) look first at the case in which the coordinates system is the initial page coordinate system; (2) reduce the general case to that one.

Suppose for the moment that we are working with page coordinates with the origin at lower right and units that are points. In these circumstances, we divide the problem into two cases: (a) that in which the line is "essentially" horizontal; (b) that in which it is "essentially" vertical. We could in fact divide the cases into truly vertical (where $B = 0$) and the rest, but for technical reasons, having B near 0 is almost as bad as having it actually equal to 0. In this scheme, we consider a line essentially horizontal if its slope lies between -1 and 1; otherwise it is considered essentially vertical. In other words, we think of it as essentially horizontal if it is more horizontal than vertical. Recalling that if a line has equation $Ax + By + C = 0$ then the direction $[A, B]$ is perpendicular to that line, we have the following criterion:

■ *The line $Ax + By + C = 0$ will be considered "essentially horizontal" if $|A| \leq |B|$, otherwise it is considered "essentially vertical."*

Recall that our coordinate system is in points. The left-hand side of the page is therefore at $x_{\text{left}} = 0$, and the right one at $x_{\text{right}} = 72 \cdot 8.5 = 612$. The point is that *an essentially horizontal line is guaranteed to intercept both of the lines* $x = x_{\text{left}}$ *and* $x = x_{\text{right}}$. Why? Because A and B cannot both be 0 and $|A| \leq |B|$ for a horizontal line, we must have $B \neq 0$ as well. Therefore, we can solve to get $y = (-C - Ax)/B$, where we choose x to be in turn x_{left} and x_{right}. In this case, we choose these intercepts for P and Q. It may happen that P or Q is not on the edge of the page,

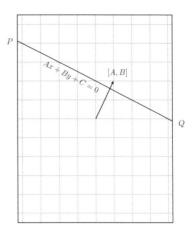

and it may even happen that the line segment PQ is totally invisible, but this doesn't matter. What does matter is that the segment PQ is guaranteed to contain all of the visible part of the line. Similarly, an essentially vertical line must intercept the lines across the top and bottom of the page, and in this case these intercepts will be P and Q.

Thus, we must design a procedure in PostScript that does one thing for essentially horizontal lines and another for essentially vertical ones. We need to use a **test** together with a **conditional** in our procedure.

A test is a command sequence in PostScript that returns one of the **Boolean values** true or false on the stack. We will find several useful: le, lt, ge, gt, eq, ne, which stand for $\leq$, $<$, $\geq$, $>$, $=$, and $\neq$, respectively. They are used backwards, of course. For example, the command sequence

```
a b lt
```

will put true on the stack if $a < b$; otherwise, false will be put on the stack.

Here is a sample from a Ghostscript session:

```
1 2 gt =
false
2 1 gt =
true
```

A **conditional** is a command sequence that does one thing in some circumstances and something else in others. The most commonly used form of a conditional is this:

```
boolean
{ ... }
{ ... }
ifelse
```

That is, we include a few commands to perform a test of some kind, following the test with two procedures and the command ifelse. If the result of the test is true, the first procedure is performed; otherwise, the second. Recall that a procedure in PostScript is any sequence of commands entered on the stack surrounded by { and }.

A slightly simpler form is also possible:

```
boolean
{ ... }
if
```

This performs the procedure if the boolean is true and otherwise does nothing.

To apply if or ifelse, normally you apply a test of some kind. You can combine tests with and, not, or.

We now have just about everything we need to write the procedure mkline. Actually, for reasons that will become clear in a moment, we make up a procedure called segment-page that instead of building the line returns (i.e., leaves on the stack) the two endpoints P and Q as a pair of arrays of two numbers. We need to recall that x abs returns the absolute value of x.

```
% [A B C] on stack
/segment-page { 1 dict begin
aload pop
/C exch def
/B exch def
/A exch def

A abs B abs le
{
  /xleft 0 def
  /xright 612 def
  /yleft C A xleft mul add B div neg def
  % y = -C - Ax / B
  /yright ... def
  [xleft yleft]
```

```
        [xright yright]
    }{
        . . .
    } ifelse
    end } def
```

I have left a few blank spots – on purpose.

EXERCISE 4.8. *Fill in the . . . to get a working procedure. Demonstrate it with a few samples.*

EXERCISE 4.9. *Show how to use this procedure in a program that draws the line* $113x + 141y - 300$ *in page coordinates.*

Now we cease to assume that we are dealing with page coordinates. We would like to make up a similar procedure that works no matter what the user coordinates are. So we are looking for a procedure with a single array argument [A B C] that builds in the current coordinate system, no matter what it may be, a line segment including all of the line that's visible on a page. This takes place in three stages: (1) We find the equation of the line in page coordinates; (2) we apply segment-page to find points containing the segment we want to draw in page coordinates; (3) we transform these points back into user coordinates.

Here is the final routine we want:

```
% [A B C] on stack
/mkline { 1 dict begin
  aload pop
  /C exch def
  /B exch def
  /A exch def
   % T = page to user matrix
  /T
    matrix defaultmatrix
    matrix currentmatrix
    matrix invertmatrix
    matrix concatmatrix
  def
   % get line in page coordinates
```

```
    [
      A T 0 get mul
      B T 1 get mul add

      A T 2 get mul
      B T 3 get mul add

      T 4 get A mul
      T 5 get B mul add
      C add
    ]
     % find P, Q
    segment-page
     % build the line
    aload pop T transform moveto
    aload pop T transform lineto
  end } def
```

EXERCISE 4.10. *Finish the unfinished procedures you need and assemble all the pieces into a pair of procedures that will include this main procedure. Exhibit some examples of how things work.*

4.7   ANOTHER WAY TO DRAW LINES

There is another way to solve the problem posed in this chapter and that is to find exactly where the line enters and exits the page. This reduces to the following problem: *Suppose (x_0, y_0) and (x_1, y_1) are the lower left and upper right corners of a rectangle and $Ax + By + C = 0$ is the equation of a line. How can you find the points where the line intersects the boundary of the rectangle?*

There are many possibilities, of course. The line could intersect in 0, 1, or 2 points, and in the course of solving the problem you must decide which. We'll deal with a related problem in a more organized fashion later on when we look at the Hodgman–Sutherland algorithm, but the problem here, although somewhat related, is simpler.

We want to define a procedure with three arguments – (x_0, y_0), (x_1, y_1) (corners of the rectangle) and $[A, B, C]$ – that returns an array of 0, 1, or 2 points of intersection. Details will be left as an exercise, but I want to explain here some of the features of the procedure that will be applicable in other situations too.

Let's look at an explicit example to see how things are going to go. Suppose we ask whether the line $f(x, y) = 2x + 3y - 1 = 0$ intersects the frame of the unit square and, if so, where it intersects. The thing to keep in mind here is that the computer is, of course, blind. It can't see anything, and so how is it going to figure out where intersections occur? The fundamental criterion is this: *If P and Q are two points on opposite sides of the line, then $f(P)$ and $f(Q)$ will have different signs.* In other words, we evaluate $f(x, y)$ at each of the corners of the square, and if we find sides where the signs at the ends are different, we have an intersection. Here is what happens in this case:

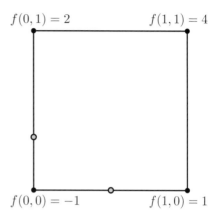

There are two sides where the sign of $f(x)$ is different at the end points. At the bottom the range is -1 to 1, and $f(x)$ will be 0 at the midpoint. At the left the range is -1 to 2, which tells us that the intersection point is $1/3$ of the way from bottom to top.

What happens in the procedure? Start at one corner, the lower left for instance, and then "walk around" the sides of the rectangle, checking to see if the line $Ax + By + C$ intersects that side. There are a few subtle points to be taken into account.

One subtle thing is to be careful exactly what a **side** is. As we walk around the sides, we traverse them in a particular direction; the sides are **oriented** segments. It turns out to be best to define a side of the rectangle to include its endpoint but exclude its beginning. Thus, a side looks like this metaphorically:

Before I start with details, let's think about the possibilities. (1) There could of course be no point of intersection at all. (2) There could be a single intersection, which could be either (a) in the middle or (b) at the endpoint. (3) The whole side could be part of the line.

How can we distinguish these cases? Basically, we do so by checking the sign of

$Ax + By + C$ at the two ends of the side. Let

$$F_P = Ax_P + By_P + C, \qquad F_Q = Ax_Q + By_Q + C.$$

Here is a complete logical breakdown:

- If $F_P < 0$ and $F_Q > 0$, there is a single interior point of intersection.
- Same conclusion if $F_P > 0$ and $F_Q < 0$.
- If $F_P < 0$ and $F_Q = 0$, then the endpoint Q is a point of intersection.
- The same conclusion obtains if $F_P > 0$ and $F_Q = 0$.
- If $F_P = 0$ and $F_Q = 0$, the side is contained in the line.
- In all other cases there is no intersection.

One question is how to deal with the case in which the side is contained in the line. It turns out that there is one best thing to do, and that is to treat it exactly the same as other cases in which Q is on the line. This will work because, on the previous side the point P will be counted and the procedure will return both P and Q, which is quite reasonable.

The breakdown can be summarized: (1) if $F_P F_Q < 0$, then there is an interior point; (2) if $F_Q = 0$, then Q is the point of intersection; (3) otherwise, there is no intersection.

Another problem is mathematical. In the case of an interior point of intersection, how is it calculated? The function $Ax + By + C$ is equal to $F(P)$ at P, 0 at the intersection point and $F(Q)$ at Q. The increase of F across the segment is linear, and thus

$$F((1 - t)P + tQ) = (1 - t)F(P) + tF(Q) = F(P) + t(F(Q) - F(P)),$$

and if we want to get $F((1 - t)P + tQ) = 0$ we set

$$t = \frac{F(P)}{F(P) - F(Q)}.$$

I can say some more about what the procedure does. After defining some variables, it puts a [down on the stack and then looks at each side in turn. If there is a point of intersection (x, y), the procedure puts it on the stack as an array of two points. Otherwise, it does nothing. As the procedure exits, it puts] on the stack. What's returned on the stack will be thus of the form [], [[..]], or [[..][..]]. I can offer one hint, too, for efficiency: start off with P equal to the corner point and calculate F_P immediately. When you have looked at a side, define the new values of P and F_P to be the current values of Q and F_Q and go on to the next side.

EXERCISE 4.11. *Define the procedure in detail. Write it so that it will handle any convex closed polygon, that is to say one that bulges out, so that the intersection of a line with it is always either empty, a single point, two points, or a whole side.*

4.8 CLIPPING

It might be that we don't want to draw all of the visible line but want to allow some margins on our page. We could modify the procedure very easily to do this by changing the definitions of `xleft` and so on, but this is inelegant because it would require putting in a new procedure for every different type of margin. There is a more flexible way. There is a third command in the same family as `stroke` and `fill` called `clip`. It, too, is applied to a path just constructed. Its effect is to restrict drawing to the interior of the path. Thus,

```
newpath
72 72 moveto
540 72 lineto
540 720 lineto
72 720 lineto
closepath
clip
```

creates margins of size $1''$ on an $8.5'' \times 11''$ page in page coordinates. If you want to restrict drawing for a while and then abandon the restriction, you can enclose the relevant stuff inside `gsave` and `grestore`. The command `clip` is like `fill` in that it will automatically close a path before clipping to it, but as with `fill` it is not a good habit to rely on this. This is another example of the idea that *programs should reflect concepts*: if what you really have in mind is a closed path, close it yourself. The default closure may not be what you intend.

4.9 ORDER COUNTS

This seems like a good place to recall that the order in which a sequence of coordinate changes takes places is important. Sometimes this is a useful feature; sometimes it is just a nuisance.

I have said before that nonuniform scaling (i.e., scaling by different factors on the different axes) can have peculiar effects. This is particularly so when applying rotations. It is very important in what order rotating and scaling occur if the scaling is not uniform.

Here is what happens for each of these sequences:

```
30 rotate
1 1.5 scale

newpath
0.5 square
stroke


1 1.5 scale
30 rotate

newpath
0.5 square

stroke
```

The point to keep in mind is something I have said before: after a coordinate change is made, all further coordinate changes take place *with respect to that new system*. Rotation always preserves the curve $x^2 + y^2 = 1$. But if x and y are scaled by different factors, a circle will become an ellipse and rotation takes place *around this ellipse* instead of around a true circle. We can see what happens by doing it in stages:

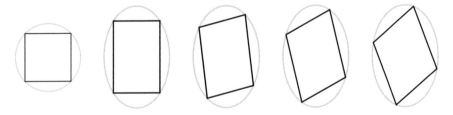

4.10   CODE

The code to draw lines can be found in `lines.inc`.

CHAPTER 5

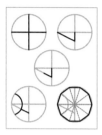

Drawing polygons: loops and arrays

We begin by learning how to draw regular polygons and then look at arbitrary polygons. Both will use loops, and arbitrary polygons will require learning about arrays.

There are several kinds of loop constructs in PostScript. Three are frequently used.

5.1 THE REPEAT LOOP

The simplest loop is the `repeat` loop. It works very directly. The basic pattern is

```
N {
...
} repeat
```

Here N is an integer. Next comes a procedure followed by the command `repeat`. The effect is very simple: the lines of the procedure are repeated N times. Of course these lines can have side effects, and so the overall complexity of the loop might not be negligible.

One natural place to use a loop in PostScript is to draw a regular N-sided polygon. This is a polygon that has N sides, all of the same length, and also possesses central symmetry around a single point. If you are drawing a regular polygon by hand, the simplest thing to do first is to draw a circle of the required size, mark N points evenly around this circle, and then connect neighbors. Here we will assume that the radius is to be 1 and that the location of the N points is fixed by assuming one of them to be $(1, 0)$.

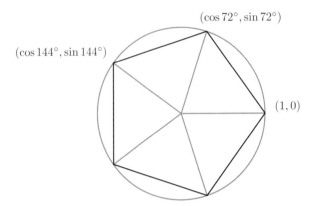

If we set $\theta = 360/N$, then the other points on the circle will be $(\cos\theta, \sin\theta)$, $(\cos 2\theta, \sin 2\theta)$, and so on. To draw the regular polygon, we first move to $(1, 0)$ and then add the lines from one vertex to the next. At the nth stage we must add a line from $(\cos(n - 1)\theta, \sin(n - 1)\theta)$ to $(\cos n\theta, \sin n\theta)$. How can we make this into a repetitive action? This can be done by using a variable to store the current angle and incrementing it by $360/N$ in each repeat. Here is a procedure that will do the job:

```
% At entrance the number of sides is on the stack
 % The effect is to build a regular polygon of N sides
/make-regular-polygon { 4 dict begin
  /N exch def
  /A 360 N div def
  1 0 moveto
  N {
    A cos A sin lineto
    /A A 360 N div add def
  } repeat
  closepath
end } def
```

In the first iteration, $A = 360/N$, in the second $A = 720/N$, and so on.

EXERCISE 5.1. *Modify this procedure to have two arguments, the first equal to the radius of the polygon. Why is it not worthwhile to add the center and the location of the initial point as arguments?*

5.2 THE FOR LOOP

Repeat loops are the simplest in PostScript. Slightly more complicated is the `for` loop. To show how it works, here is an example that draws a regular pentagon:

```
1 0 moveto
1 1 5 {
  /i exch def
  i 72 mul cos i 72 mul sin lineto
} for
closepath
```

The `for` loop has one slightly tricky feature that requires the line `/i exch def`. The structure of the `for` loop is this:

```
s h N {
  ...
} for
```

This loop involves a "hidden" and nameless variable that starts with a value of s, increments itself by h each time the procedure is performed, and stops after doing the last loop where this variable is equal to or less than N. This hidden (or implicit) variable is put on the stack just before each repetition of the procedure. The line `/i exch def` behaves just like the similar lines in procedures – it takes that hidden variable off the stack and assigns it to the named variable i. It is not necessary to do this, but you must do *something* with that number on the stack because otherwise it will just accumulate there, causing eventual if not immediate trouble. If you don't need to use the loop variable but just want to get rid of it, use the command pop, which just removes the top item from the stack.

Incidentally, it is safer to use only integer variables in the initial part of a `for` loop because otherwise rounding errors may cause a last loop to be missed or an extra one to be done.

EXERCISE 5.2. *Make up a procedure* polygon *just like the one in the first section but using a* for *loop instead of a* repeat *loop.*

EXERCISE 5.3. *Write a complete PostScript program that makes your own graph paper. There should be light gray lines 1 mm apart and heavier gray ones 1 cm apart, and the axes should be done in black. The center of the axes should be at the center of the page. Fill as much of the page as you can with the grid.*

5.3   THE LOOP LOOP

The third kind of loop is the most complicated but also the most versatile. It operates somewhat like a `while` loop in other languages but with a slight extra complication.

```
1 0 moveto
/A 72 def
{ A cos A sin lineto
  /A A 72 add def
  A 360 gt { exit } if
} loop
closepath
```

The complication is that you *must* test a condition in the loop and explicitly force an exit if it is not satisfied. Otherwise, you will loop forever. If you put in your condition at the beginning of the loop, you have the equivalent of a `while` loop, whereas if at the end a `do ... while` loop. Thus, the commands `loop` and `exit` should almost always be used together. Exits can be put into any loop to break out of it under exceptional conditions.

5.4   GRAPHING FUNCTIONS

Function graphs or parametrized curves can de done easily with simple loops, although we will see a more sophisticated way to do them in the next chapter. Here is sample code to draw a graph of $y = x^2$ from -1 to 1:

```
/N 100 def
/x -1 def
/dx 2 N div def

/f {
  dup mul
} def

newpath
x dup f moveto
N {
  /x x dx add def
  x dup f lineto
} repeat
stroke
```

| 5.5 | **GENERAL POLYGONS** |

Polygons don't have to be regular. In general a polygon is essentially a sequence of points P_0, P_1, ..., P_{n-1} called its **vertices**. The edges of the polygon are the line segments connecting the successive vertices. We impose a convention here: a point will be an array of two numbers $[x\ y]$ and a polygon will be an array of points $[P_0\ P_2\ \ldots\ P_{n-1}]$. We now want to define a procedure that has an array like this as a single argument and builds the polygon from that array by making line segments along its edges.

There are a few things you have to know about arrays in PostScript to make this work (and they are just about all you have to know):

(1) The numbering of items in an array starts at 0;

(2) if a is an array, then a length returns the number of items in the array;

(3) if a is an array, then a i get puts the ith item on the stack;

(4) you create an array on the stack by entering [, a few items, then];

```
% argument: array of points
 % builds the corresponding polygon
/make-polygon { 3 dict begin
/a exch def
/n a length def
n 1 gt {
  a 0 get 0 get
  a 0 get 1 get
  moveto
  1 1 n 1 sub {
    /i exch def
    a i get 0 get
    a i get 1 get
    lineto
  } for
} if
end } def
```

This procedure starts out by defining the local variable a to be the array on the stack, which is its argument. Then it defines n to be the number of items in a. If $n \leq 1$ there is nothing to be done at all. If $n > 1$, we move to the first point in the array and then draw $n - 1$ line segments. Because there are n points in the array, we draw $n - 1$ segments, and the last point is P_{n-1}. Note also that, since the ith item in the array is a point P_i, which is itself an array of two items, we must "get" its elements to make a line. If $P = [x \; y]$, then P 0 get P 1 get puts x y on the stack.

There is another way to unload the items in an array onto the stack: the sequence P aload puts all the entries of P onto the stack together with the array P itself at the top. The sequence P aload pop thus puts all the entries on the stack in order. This is simpler and more efficient than getting the items one by one.

Note also that, if we want a closed polygon, we must add closepath outside the procedure. There is no requirement that the first and last points of the polygon be the same.

There is one more important thing to know about arrays. Normally, you build one by entering any sequence of items in between square brackets [and], separated by space, possibly on separate lines. An array can be any sequence of items, not necessarily all of the same kind. The following is a legitimate use of make-polygon to draw a pentagon:

```
newpath
[
  [1 0]
  [72 cos 72 sin]
  [144 cos 144 sin]
  [216 cos 216 sin]
  [288 cos 288 sin]
]
make-polygon
closepath
stroke
```

EXERCISE 5.4. *Use loops and* make-polygon *to draw the American flag in color 3″ high and 5″ wide for instance. (The stars – there are 50 of them – are the interesting part.)*

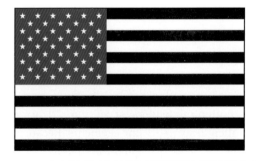

EXERCISE 5.5. *Another useful pair of commands involving arrays are* `array` *and* `put`. *The sequence* `n array` *puts on the stack an array of length n. What is in it? We find a sequence of* `null` *objects – that is, essentially faceless entities. Of course the array will be of no use until null objects are replaced by proper data. The way to do this is with the* `put` *command. The sequence* `A n x put` *sets* $A[n] = x$.

(1) Construct a procedure `reversed` *that replaces an array by the array that lists the same objects in the opposite order without changing the original array. (2) Then use* `put` *and careful stack manipulations to do this without using any variables in your procedure. Thus*

> `[0 1 2 3] reversed`

should return `[3 2 1 0]`.

5.6 CLIPPING POLYGONS

In this section, now that we are equipped with loops and arrays, we take up a slight generalization of the problem we began with in the last chapter. We will also see a new kind of loop.

Here is the new problem:

- We are given a closed planar polygonal path γ together with a line $Ax + By + C = 0$. We want to replace γ by its intersection with the half plane $f(x, y) = Ax + By + C \leq 0$.

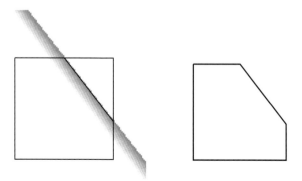

We are going to see here a simple but elegant way of solving these problems called the **Hodgman–Sutherland algorithm** after its inventors. We may as well assume the path to be oriented. The solution to the problem reduces to one basic idea: if the path γ exits the half plane at a point P and next crosses back at Q, we want to replace that part of γ between P and Q by the straight line PQ.

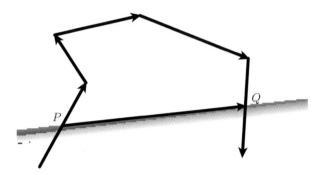

We want to design a procedure, which I'll call hodgman-sutherland, that has the closed polygon γ and the line $Ax + By + C = 0$ as arguments and returns on the stack a new polygon obtained from γ by cutting off the part in the region $Ax + By + C > 0$. The polygon γ will be represented by the array of its vertices and the line by the array $\ell = (A, B, C)$. Let

$$\langle \ell, P \rangle = Ax + By + C$$

if $P = (x, y)$.

The procedure looks in turn at each edge of the polygon in the order determined by the array. It starts with the edge P_{n-1}, P_0, which is a convenient trick in such situations. As we proceed, we are going to build up the replacement polygon by adding points to it. Suppose we are looking at an edge PQ. What we do will depend on circumstances. Roughly put,

(1) If $\langle \ell, P \rangle \leq 0$ and $\langle \ell, Q \rangle \leq 0$ (both P and Q inside the half plane determined by ℓ), we add Q to the new polygon;

(2) if $\langle \ell, P \rangle < 0$ but $\langle \ell, Q \rangle > 0$ (P inside, Q outside), we add the intersection $PQ \cap \ell$ of the segment PQ with ℓ, which is

$$\frac{\langle \ell, Q \rangle P - \langle \ell, P \rangle Q}{\langle \ell, Q \rangle - \langle \ell, P \rangle},$$

to the new polygon;

(3) if $\langle \ell, P \rangle = 0$ but $\langle \ell, Q \rangle > 0$, we do nothing;

(4) if $\langle \ell, P \rangle > 0$ but $\langle \ell, Q \rangle \leq 0$ (P outside, Q inside), then we add both $PQ \cap \ell$ and Q unless they are the same, in which case we just add one;

(5) if both P and Q are outside, we do nothing.

This process is very much like that performed in Chapter 4 to find the intersection of a line and a rectangle, and one convention is certainly the same – an edge does not really contain its starting point.

In certain singular cases, one of the vertices lies on ℓ and no new point is calculated.

One peculiar aspect of the process is that if the line crosses and recrosses several times, it still returns a single polygon.

In writing the procedure to do this, we are going to use the `forall` loop. It is used like this:

```
a {
   ...
} forall
```

where a is an array. The procedure { ... } is called once for each element of the array a with that element on top of the stack. It acts much like the `for` loop, and in fact some `for` loops can be simulated with an equivalent `forall` loop by putting an appropriate array on the stack.

In the program excerpt below there are a few things to notice in addition to the use of `forall`. One is that, for efficiency's sake, the way in which a local dictionary is used is somewhat different from what we have encountered previously. I have defined a procedure `evaluate` that calculates $Ax + By + C$ given [A B C] and [x y]. If I were following the pattern recommended earlier, this procedure would set up its own local dictionary on each call to it. But setting up a dictionary is inefficient, and `evaluate` is called several times in the main procedure here. So I use no dictionary but rely entirely on stack operations. The new stack operation used here is `roll`. It has two arguments, n and i, shifting the top n elements on the stack cyclically up by i – that is, it rolls the top n elements of the stack. If the stack is currently

$$x_4 \; x_3 \; x_2 \; x_1 \; x_0 \quad \text{(bottom to top)}$$

then the sequence 5 2 `roll` changes it to

$$x_1 \; x_0 \; x_4 \; x_3 \; x_2.$$

```
         % x y [A B C] => Ax + By + C
/evaluate {   % x y [A B C]
   aload pop   % x y A B C
   5 1 roll    % C x y A B
   3 2 roll    % C x A B y
   mul         % C x A By
   3 1 roll    % C By x A
   mul         % C By Ax
   add add     % Ax+By+C
} def
```

Another thing to notice is that the data we are given are the vertices of the polygon, but what we really want to do is look at its edges or pairs of successive vertices. So we use two variables P and Q and start with $P = P_{n-1}$. In looping through P_0, $P_1, \ldots$ we are therefore looping through edges $P_{n-1} P_0$, $P_0 P_1, \ldots$

```
% arguments: polygon [A B C]
% returns: closure of polygon truncated to Ax+By+C <= 0
/hodgman-sutherland { 4 dict begin
/f exch def
/p exch def
/n p length def
 % P = p[n-1] to start
/P p n 1 sub get def
/d P length 1 sub def
/fP P aload pop f evaluate def
[
  p {
    /Q exch def
    /fQ Q aload pop f evaluate def
    fP 0 le {
      fQ 0 le {
        % P <= 0, Q <= 0: add Q
        Q
      }{
        % P <= 0, Q > 0
        fP 0 lt {
          % if P < 0, add intersection
          /QP fQ fP sub def
          [
            fQ P 0 get mul fP Q 0 get mul sub QP div
            fQ P 1 get mul fP Q 1 get mul sub QP div
          ]
        } if
      } ifelse
    }{
      % P > 0
      fQ 0 le {
        % P > 0, Q <= 0: if fQ < 0, add intersection;
        % add Q in any case
        fQ 0 lt {
```

```
            /QP fQ fP sub def
            [
              fQ P 0 get mul fP Q 0 get mul sub QP div
              fQ P 1 get mul fP Q 1 get mul sub QP div
            ]
          } if
          Q
        } if
        % else P > 0, Q > 0: do nothing
      } ifelse
      /P Q def
      /fP fQ def
    } forall
  ]
end } def
```

EXERCISE 5.6. *If a path goes exactly to a line and then retreats, the code above will include in the new path just the single point of contact. Redesign the procedure so as to put two copies of that point in the new path. It is often useful to have a path cross a line in an even number of points.*

5.7 CODE

There is a sample function graph in `function-graph.ps` and code for polygon clipping in dimensions three as well as two in `hodgman-sutherland.inc`.

CHAPTER 6

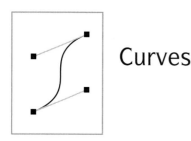

Curves

So far, the only paths we have learned to draw in PostScript are sequences of line segments. It is possible to assemble a good approximation of just about any curve by a large number of segments, but there are more elegant and efficient ways involving **Bézier curves**.

6.1 ARCS

The simplest curves are circles. There are two special commands to draw circles and pieces of circles. The sequence

```
0 0 10 47 67 arc
```

will add to the current path the short arc of a circle of radius 10, centerd at the origin, between arguments 47° and 67°. If arcn is used, it will draw the clockwise arc around the long way instead (arcn for **arc n**egative).

It would be a good idea to investigate here what "adding to the current path" means because it is behavior many of our later procedures will imitate. Here are two short sketches that should illustrate how it works. In the first, we draw a line and then continue drawing an arc. The default behavior is for an arc to continue the current path in this way – to add a line from the last point of the previous path to the first point of the arc. Sometimes this is not what one wants or expects, in which case it is necessary to add a moveto to break up the path, as in the second figure.

```
newpath
0 0 moveto
1 0 lineto
0 0 1 45 90 arc
stroke
```

```
newpath
0 0 moveto
1 0 lineto
45 cos 45 sin moveto
0 0 1 45 90 arc
stroke
```

Another curiousity of arc is that in a coordinate system in which y-units are distinct from x-units it produces an ellipse. In other words, it always draws the locus of an equation $(x - a)^2 + (y - b)^2 = r^2$ in user coordinates (x, y). If the axes are not perpendicular or the x and y units are different, it will look like an ellipse.

6.2 FANCIER CURVES

Lines and arcs of circles make up a very limited repertoire. PostScript allows a third method to build paths that is much more versatile. In creating complicated paths such as the outlines of characters in a font, this third method is indispensable.

Conceptually, the simplest way to draw even a complicated curve is by drawing a sequence of line segments – that is, making a **polygonal approximation** to it – but this usually requires a very large number of segments to be at all acceptable. It also suffers from the handicap that it is not very scalable – that is, even if a collection of segments looks smooth at one scale, it may not look good at another. Here, for example, is a portion of the graph of $y = x^4$ drawn with eight linear segments.

The overall shape is not too bad, but the breaks are quite visible. You can certainly improve the quality of the curve by using more segments, but then the number of segments required to satisfy the eye changes with the scale used in representing the curve. One trouble is that the human eye can easily  perceive that the directions vary discontinuously in this figure. This is not at all something to be taken for granted. The more we learn about vision in nature, the more we learn that most features like this depend on sophisticated image processing. In contrast, the human eye apparently has trouble perceiving discontinuities in curvature.

In any event, it is better, if possible, to produce smooth curves – at least as smooth as the physical device at hand will allow. PostScript does this by approximating segments of a curve by **Bézier cubic curves**. This allows us to have the tangent direction of a curve vary continuously as well.

The Bézier curve is the last major ingredient of PostScript to be encountered. In the rest of this book, we will learn how to manipulate and combine the basic tools we have already been introduced to.

6.3   BÉZIER CURVES

In PostScript, to add a curved path to a path already begun, you put in a command sequence like

```
1 1
2 1
3 0
curveto
```

This makes a curve starting at the **current point** P_0, ending at $P_3 = (3, 0)$, and in between following a path **controlled** by the intermediate points $P_1 = (1, 1)$ and $P_2 = (2, 1)$. If there is no current point, a moveto command should precede this. Thus,

```
0 0 moveto
1 1
2 1
3 0
curveto
```

would make a complete curve starting at $(0, 0)$.

In short, curveto behaves very much like lineto but depends on a larger set of points. At any rate, what we get is this (where the four relevant points are marked):

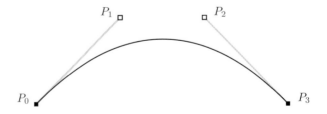

In this book I usually call P_0 and P_3 the **end points** and P_1 and P_2 the **control points** of the curve. Occasionally, I just call all of them control points, which is more standard terminology. Even from this single picture you will see that the effect of the control points on the shape of the curve is not so simple. To draw curves efficiently and well, we have to understand this matter much better. We will see in Section 6.5 the exact mathematics of what is going on, but right now I simply exhibit several examples.

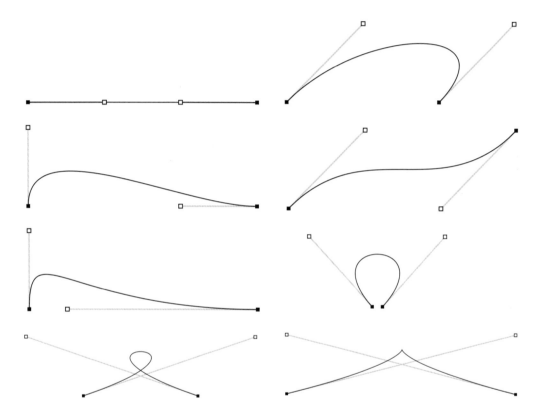

You should be able to see from these examples that the use of control points to specify curves becomes intuitive with experience. The following facts may give a rough feeling for how things go.

(1) *The path starts at P_0 and ends at P_3.*

(2) *When the curve starts out from P_0, it is heading straight for P_1.*

(3) *Similarly, when it arrives at P_3, it is coming from the direction of P_2.*

(4) *The longer the line from P_0 to P_1, the tighter the curve sticks to that line when it starts out from P_0. The behavior of P_2 and P_3 is similar.*

There is another fact that is somewhat less apparent.

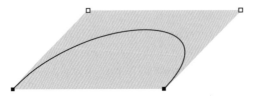

(5) *If we wrap up the four points P_i in a quadrilateral box, then the whole curve is contained inside that box.*

The intuitive picture of a Bézier curve conceives of it as a path followed by a particle in motion between certain times. The vector from P_0 to P_1 is proportional to the velocity of the particle as it starts out from P_0, and the vector from P_2 to P_3 is proportional to its velocity when it arrives at P_3. Another way of putting this is that, roughly speaking, the control points are a convenient way to encode the initial and final velocities in geometric data. This explains properties (2), (3), and (4). Property (5) is implied by the fact that any point on the curve is some kind of weighted average of the four points P_i, as we will see in Section 6.8.

Curves drawn by using control points in this way are called **Bézier curves** after the twentieth-century French automobile designer Pierre Bézier, who was one of the very first to use them extensively in computer graphics even though their use in mathematics under the name of **cubic interpolation curves** is much older.

One natural feature of Bézier curves described by control points is that they are stable under arbitrary affine transformations – that is, the affine transformation of a Bézier curve is the Bézier curve defined by the affine transformations of its control points. This is often an extremely useful property to keep in mind.

EXERCISE 6.1. *Write a PostScript procedure* `pixelcurve` *with arguments 4 arrays P_0, P_1, P_2, P_3 of size 2 with the effect of drawing the corresponding Bézier curve, including also black pixels of width 0.05″, at each of these points.*

6.4   HOW TO USE BÉZIER CURVES

In this section a recipe for using Bézier curves to draw very general curves is introduced. In the next section, this recipe will be justified. To make the recipe plausible, we begin by looking at the problem of how to approximate a given curve by polygons.

The first question we must answer, however, is more fundamental: *How are curves to be described in the first place?* In this book the answer will usually be in terms of a **parametrization**. Recall that a **parametrized curve** is a map from points of the real line to points in the plane – that is, to values of t in a selected range we associate points $(x(t), y(t))$ in the plane. It often helps one's intuition to think of the **parameter** t as time, and thus as time proceeds we move along the curve from one point to another. In this scheme, with a parametrization $P(t)$, the **velocity** vector at time t is the limit of average velocities over smaller and smaller intervals of time $(t, t + h)$:

$$V(t) = P'(t) = \lim_{h \to 0} \frac{P(t + h) - P(t)}{h} = [x'(t), y'(t)].$$

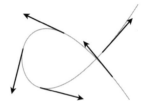

The direction of the velocity vector is tangent to the curve, and its magnitude is determined by the speed of motion along the curve.

Example. The unit circle with center at the origin has parametrization $t \mapsto (\cos t, \sin t)$.

Example. If $f(x)$ is a function of one variable x, its graph has the parametrization $t \mapsto (t, f(t))$.

In other words, a parametrization is essentially just a pair of functions $(x(t), y(t))$ of a single variable, which is called the parameter. The parameter often has geometric significance. For example, in the parametrization $t \mapsto (\cos t, \sin t)$ of the unit circle it is the angle at the origin between the positive x-axis and the radius to the point on the circle.

Example. Besides the standard parametrization of the circle there is another interesting one. If ℓ is any line through the point $(1, 0)$ other than the vertical line $x = 1$, it will intersect the circle at exactly one other point on the circle.

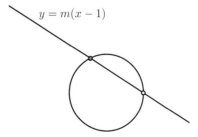

The equation of such a line will be $y = m(x - 1) = mx - m$, where m is its slope.

The condition that a point $(x, mx - m)$ lies on the circle is

$$x^2 + y^2 = 1$$
$$= x^2 + m^2(x - 1)^2$$
$$= x^2(1 + m^2) - 2m^2 x + m^2$$

$$x^2 - 2\frac{m^2}{m^2 + 1}x + \frac{m^2 - 1}{m^2 + 1} = 0$$

$$(x - 1)\left(x - \frac{m^2 - 1}{m^2 + 1}\right) = 0,$$

and thus

$$x = \frac{m^2 - 1}{m^2 + 1}, \quad y = \frac{-2m}{m^2 + 1}.$$

As m varies from $-\infty$ to ∞, the point (x, y) traverses the whole circle except the point $(1, 0)$. Thus, m is a parameter and

$$m \longmapsto \left(\frac{m^2 - 1}{m^2 + 1}, \frac{-2m}{m^2 + 1}\right),$$

which is a kind of parametrization of the unit circle. This has historical significance. If we set $m = p/q$ to be a fraction, the point (x, y) will be a point on the unit circle with rational coordinates such as $(a/c, b/c)$ with $(a/c)^2 + (b/c)^2 = 1$. If we clear denominators, we obtain a set of three integers a, b, c with $a^2 + b^2 = c^2$. Such a set is called a **Pythagorean triple**. There is evidence that this construction was known to the Babylonians in about 1800 B.C.

EXERCISE 6.2. *Use this idea to find the smallest several Pythagorean triples.*

Example. There are two common ways to specify a curve in the plane. The first is a parametrization. The second is an equation relating x and y. An example is the oval

$$x^4 + y^4 = 1.$$

This is not the graph of a function, and it has no obvious single parametrization. We can solve the equation $x^4 + y^4 = 1$ to get

$$y = \sqrt[4]{1 - x^4},$$

which gives the top half of our oval, and obtain the bottom half similarly. Neither half is yet the graph of a good function, however, because both have infinite slope at $x = \pm 1$. We can nevertheless restrict the range of x away from ± 1 – For instance to $[-\sqrt[4]{1/2}, \sqrt[4]{1/2}]$. We can then turn the curve sideways and now solve for x in terms of y to write the rest as a graph rotated $90°$. To summarize, we can at least express this curve as the union of four separate pieces, each of which we can deal with.

EXERCISE 6.3. *Find a parametrization of this oval by drawing inside it a circle and taking as the point corresponding to t the point of intersection of the oval with the ray from the origin at angle t.*

EXERCISE 6.4. *Sketch the curve $y^2 = x^2(x + 1)$ by hand in the region $|x| \leq 3$, $|y| \leq 3$. Find a parametrization of this curve by using the fact that the line $y = mx$ will intersect it at exactly one point other than the origin. Write down this parametrization. Use it to redo your sketch in PostScript in any way that looks convincing in order to check your drawing.*

With this understanding of how a curve is given to us, the question we are now confronted with is this:

- ■ *Given a parametrization $t \mapsto P(t)$ of a curve in the plane, how do we draw part of it using Bézier curves?*

If we were to try to draw it using linear segments, the answer would go like this: Suppose we want to draw the part between given values t_0 and t_1 of t. We divide the interval $[t_0, t_1]$ into n smaller intervals $[t_0 + ih, t_0 + (i + 1)h]$ and then draw lines $P(t_0)P(t_0 + h)$, $P(t_0 + h)P(t_0 + 2h)$, $P(t_0 + 2h)P(t_0 + 3h)$, and so on. Here $h = (t_1 - t_0)/n$. If we choose n large enough, we expect the series of linear segments to approximate the curve reasonably well.

To use Bézier curves, we will follow the roughly the same plan: chop the curve up into smaller pieces and on each small piece attempt to approximate the curve by a single Bézier curve. To do that, the essential problem we face is this: *Suppose we are given two values of the parameter t, which we may as well assume to be t_0 and t_1, and which we assume not to be too far apart. How do we approximate by a single Bézier curve the part of the curve parametrized by the range $[t_0, t_1]$?*

Calculating the end points is no problem. But how do we obtain the two interior control points? Since they have something to do with the directions of the curve at the end points, we expect to use the values of the velocity vector at the endpoints.

The exact recipe is this. Start by setting

$$P_0 = (x(t_0), y(t_0))$$
$$P_3 = (x(t_1), y(t_1)).$$

These are the end points of our small Bézier curve. Then set

$$\Delta t = t_1 - t_0$$
$$P_1 = P_0 + (\Delta t/3)P'(t_0)$$
$$P_2 = P_3 - (\Delta t/3)P'(t_1)$$

to get the control points.

Example. Let's draw the graph of the parabola $y = x^2$ for x in $[-1, 1]$. It turns out that a single Bézier curve will make a perfect fit over the whole range. Here the parametrization is $P(t) = (t, t^2)$, $P'(t) = [1, 2t]$.

$$t_0 = -1$$
$$t_1 = 1$$
$$\Delta t = 2$$
$$P_0 = (-1, 1)$$
$$P_1 = (1, 1)$$
$$P'(-1) = (1, -2)$$
$$P'(1) = (1, 2)$$
$$P_1 = P_0 + (2/3)P'(t_0)$$
$$= (-1/3, -1/3)$$
$$P_2 = (1/3, -1/3)$$

Example. Let's draw the graph of $y = x^4$ for $x = -1$ to $x = 1$. Here $P(t) = (t, t^4)$, $P'(t) = (1, 4t^3)$. We will do this with 1, 2, and 4 segments in turn.

(a) One segment $[-1, 1]$. We have this table with the control points interpolated.

x	y	x'	y'
-1.0000	1.0000	1.0	-4.0
-0.3333	-1.6667		
0.3333	-1.6667		
1.0000	1.0000	1.0	4.0

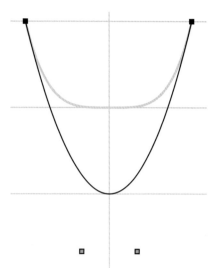

The approximation is foul, droopy.

(b) Two segments $[-1, 0]$ and $[0, 1]$.

x	y	x'	y'
−1.0000	1.0000	1.0	−4.0
−0.6667	−0.3333		
−0.3333	0.0000		
0.0000	0.0000	1.0	0.0
0.3333	0.0000		
0.6667	−0.3333		
1.0000	1.0000	1.0	4.0

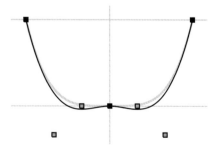

Somewhat better.

(c) Four segments $[-1.0, -0.5]$, $[-0.5, 0.0]$, $[0.0, 0.5]$, $[0.5, 1.0]$.

x	y	x'	y'
−1.0000	1.0000	1.0	−4.0
−0.8333	0.3333		
−0.6667	0.1458		
−0.5000	0.0625	1.0	−0.5
−0.3333	−0.0208		
−0.1667	0.0000		
0.0000	0.0000	1.0	0.0
0.1667	0.0000		
0.3333	−0.0208		
0.5000	0.0625	1.0	0.5
0.6667	0.1458		
0.8333	0.3333		
1.0000	1.0000	1.0	4.0

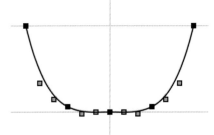

It is almost indistinguishable from the true graph. It is perhaps only when you see where the control points lie that you notice the slight rise in the middle.

EXERCISE 6.5. *In many situations, drawing a parametrized path by Bézier curves by using the velocity vector to produce control points is more trouble than it's worth. This is true even if the procedure is to be automated somewhat as explained in the next chapter, since calculating the velocity can be quite messy. One situation in Bézier plotting is definitely the method of choice, however, and that is when the path is given by a path integral. The Cornu spiral, for example, is the path in the complex plane defined by*

$$C(t) = \int_0^t e^{-is^2} \, ds$$

as t ranges from −∞ to ∞. In this case, C(t) can only be approximated incrementally by numerical methods such as Simpson's rule, but the velocity C'(t) comes out of the calculation at no extra cost since it is just the integrand. Furthermore, evaluating C(t) will be expensive in effort since each step of the approximation involves some work, and so the fewer steps taken the better.

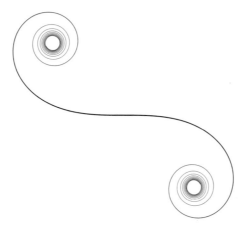

Plot the Cornu spiral, which is shown above, using Bézier curves. One subtle point in this figure is that the thickness of the path decreases as the curve spirals farther in because the spirals would otherwise be clotted.

EXERCISE 6.6. *The remarks in the previous exercise are just as valid for the plots of first-order differential equations in the plane by numerical methods. Plot, using Bézier curves, the trajectories of*

$$\begin{bmatrix} x' \\ y' \end{bmatrix} = \begin{bmatrix} -1 & -1 \\ 1 & -1 \end{bmatrix} \begin{bmatrix} x \\ y \end{bmatrix}$$

starting at a few uniformly distributed points around the unit circle.

6.5 THE MATHEMATICS OF BÉZIER CURVES

The mathematical problem we are looking at in drawing good curves in computer graphics is that of approximating an arbitrary parametrized path $t \mapsto (x(t), y(t))$ by a simpler one. If we are approximating a path by line segments, for example, then we are replacing various pieces of the curve between points $P_0 = P(t_0)$ and $P_1 = P(t_1)$ by a linearly parametrized path

$$t \mapsto \frac{(t_1 - t)P_0 + (t - t_0)P_1}{(t_1 - t_0)}$$

from one point to the other. This parametrization can be better understood if we write this as

$$t \mapsto (1-s)P_0 + sP_1, \quad \text{where} \quad s = \frac{t - t_0}{t_1 - t_0}.$$

With Bézier curves, we are asking for a parametrization from one point to the other with the property that its coordinates be cubic polynomials of t (instead of linear). In other words, we are looking for approximations to the coordinates of a parametrization by polynomials of degree three. We expect an approximation of degree three to be much better than a linear one.

The Bézier curve, then, is to be a parametrized path $B(t)$ from P_0 to P_3, cubic in the parameter t, and dependent in some way on the interior control points P_1 and P_2. Here it is:

$$B(t) = \frac{(t_1 - t)^3 P_0 + 3(t - t_1)^2(t - t_0)P_1 + 3(t_1 - t)(t - t_0)^2 P_2 + (t - t_0)^3 P_3}{(t_1 - t_0)^3}$$

$$= (1-s)^3 P_0 + 3(1-s)^2 s P_1 + 3(1-s)s^2 P_2 + s^3 P_3 \quad \left(s = (t - t_0)/(t_1 - t_0)\right).$$

We will justify this formula in Sections 6.8 and 6.9. The form using s is easier to calculate with than the other as well as more digestible.

It is simple to verify that

$$B(t_0) = P_0$$
$$B(t_1) = P_3.$$

We can also calculate (term by term)

$$(t_1 - t_0)^3 B'(t) = -3(t_1 - t)^2 P_0 + 3(t_1 - t)^2 P_1 - 6(t - t_0)(t_1 - t)P_1$$
$$+ 6(t - t_0)(t_1 - t)P_2 - 3(t - t_0)^2 P_2 + 3(t - t_0)^2 P_3$$

$$B'(t) = \frac{3(t_1 - t)^2(P_1 - P_0) + 6(T - t_0)(t - t_1)(P_2 - P_1) + 3(t - t_0)^2(P_3 - P_2)}{(t_1 - t_0)^3}$$

$$B'(t_0) = \frac{3(P_1 - P_0)}{t_1 - t_0}$$

$$B'(t_1) = \frac{3(P_3 - P_2)}{t_1 - t_0}.$$

These calculations verify our earlier assertions relating the control points to velocity since we can deduce from them that

$$P_1 = P_0 + \left(\frac{t_1 - t_0}{3}\right) B'(t_0)$$

$$P_2 = P_3 - \left(\frac{t_1 - t_0}{3}\right) B'(t_1).$$

6.6 QUADRATIC BÉZIER CURVES

A **quadratic Bézier curve** determined by three control points P_0, P_1, and P_2 is defined by the parametrization

$$Q(s) = (1-s)^2 P_0 + 2s(1-s)P_1 + s^2 P_2.$$

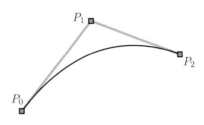

It is a degenerate case of a Bézier curve with control points P_0, $(1/3)P_0 + (2/3)P_1$, $(2/3)P_1 + (1/3)P_2$, P_2, as you can easily check. But sometimes it is easy to find control points for a quadratic curve, not so easy to find good ones for a cubic curve. One good example arises in drawing implicit curves $f(x, y) = 0$. In this case, we can often calculate the gradient vector $[\partial f/\partial x, \partial f/\partial y]$ and then that of the tangent line

$$\frac{\partial f}{\partial x}(x - x_0) + \frac{\partial f}{\partial y}(y - y_0) = 0$$

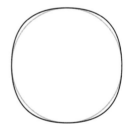

at a point (x_0, y_0) on the curve. But we can approximate the curve between two points P and Q by the quadratic Bézier curve with the intersection of the two tangent lines at P and Q as the intermediate control point. The figure below shows how the curve $x^2 + y^2 - 1 = 0$ is approximated by four quadratic curves (in red). An approximation by eight quadratic curves is just about indistinguishable from a true circle.

6.7 MATHEMATICAL MOTIVATION

In using linear or Bézier paths to do computer graphics, we are concerned with the problem of approximating the coordinate functions of an arbitrary path by polynomials of degree one or three. Considering each coordinate separately, we are led to try to approximate an arbitrary function of one variable by a polynomial of degree one or three.

The basic difference between linear approximations and cubic approximations lies in the following facts:

■ *If t_0, t_1, y_0, and y_1 are given, then there exists a unique linear function $f(t)$ such that*

$$f(t_0) = y_0$$
$$f(t_1) = y_1.$$

■ *Given t_0, t_1, y_0, y_1, v_0, v_1 there exists a unique cubic polynomial $f(t)$ such that*

$$f(t_0) = y_0$$
$$f'(t_0) = v_0$$
$$f(t_1) = y_1$$
$$f'(t_1) = v_1.$$

Roughly speaking, with linear approximations we can only get the location of end points exactly, but with cubic approximation we can get directions exact as well.

We prove here the assertion about cubic functions. If

$$f(t) = a_0 + a_1 t + a_2 t^2 + a_3 t^3,$$

then the conditions on $P(t)$ set up four equations in the four unknowns a_i, which turn out to have a unique solution (if it is assumed, of course, that $t_0 \neq t_1$). Here are the equations:

$$
\begin{array}{ccccccccc}
a_0 & + & a_1 t_0 & + & a_2 t_0^2 & + & a_3 t_0^3 & = & y_0 \\
    &   & a_1   & + & 2a_2 t_0 & + & 3a_3 t_0^2 & = & v_0 \\
a_0 & + & a_1 t_1 & + & a_2 t_1^2 & + & a_3 t_1^3 & = & y_1 \\
    &   & a_1   & + & 2a_2 t_1 & + & 3a_3 t_1^2 & = & v_1
\end{array}
$$

The coefficient matrix is

$$
\begin{bmatrix}
1 & t_0 & t_0^2 & t_0^3 \\
  & 1   & 2t_0 & 3t_0^2 \\
1 & t_1 & t_1^2 & t_1^3 \\
  & 1   & 2t_1 & 3t_1^2
\end{bmatrix}
$$

It has already been remarked that the mathematics is simplified by **normalizing** the parameter variable t, and thus instead of going from t_0 to t_1 it goes from 0 to 1. This is done by defining a new parameter variable

$$s = \frac{t - t_0}{t_1 - t_0}.$$

Note that s takes values 0 and 1 at the ends $t = t_0$ and $t = t_1$. Changing the parameter variable in this way doesn't affect the curve traversed, and simplifies algebra.

■ *Given y_0, y_1, v_0, v_1, there exists a unique cubic polynomial $f(t)$ such that*

$$f(0) = y_0$$
$$f'(0) = v_0$$
$$f(1) = y_1$$
$$f'(1) = v_1.$$

The coefficient matrix is now

$$\begin{array}{cccc} 1 & 0 & 0 & 0 \\ 0 & 1 & 0 & 0 \\ 1 & 1 & 1 & 1 \\ 0 & 1 & 2 & 3 \end{array}$$

I leave it to you as an exercise to check now by direct row reduction that the determinant is not zero, which implies that the system of four equations in four unknowns has a unique solution. Of course we know from the formula used in the previous section what the explicit formula is, but the reasoning in this section shows that this formula is the well-defined answer to a natural mathematical question.

EXERCISE 6.7. *What is the determinant of this 4×4 matrix?*

EXERCISE 6.8. *Find the coefficients a_i explicitly.*

If we put together the results of this section with those of the previous one, we have this useful characterization:

■ *Given two parameter values t_0, t_1 and four points P_0, P_1, P_2, P_3, the Bézier path $B(t)$ is the unique path $(x(t), y(t))$ with these properties:*
(1) *The coordinates are cubic as a function of t;*
(2) *$B(t_0) = P_0$, $B(t_1) = P_3$;*
(3) *$B'(t_0) = 3(P_1 - P_0)/\Delta t$ and $B'(t_1) = 3(P_3 - P_2)/\Delta t$, where $\Delta t = t_1 - t_0$.*

The new assertion here is uniqueness. Roughly, the idea is that four control points require eight numbers and that the cubic coordinate functions also require eight numbers.

6.8   WEIGHTED AVERAGES

The formula for a linear path from P_0 to P_1 is

$$P(t) = (1 - t)P_0 + tP_1$$
$$= P_0 + t(P_1 - P_0).$$

We have observed before that $P_0 + t(P_1 - P_0)$ may be seen as the point t of the way from P_0 to P_1. When $t = 0$, this gives P_0, and when $t = 1$ it gives P_1. There is also an intuitive way to understand the first formula that we have not considered so far.

Let's begin with some examples. With $t = 1/2$ we get the midpoint of the segment

$$\frac{P_0 + P_1}{2},$$

which is the average of the two. With $t = 1/3$ we get the point one-third of the way

$$\frac{2P_0 + P_1}{3},$$

which is to say that it is a **weighted average** of the endpoints with P_0 given twice as much weight as P_1.

There is a similar way to understand the formula for Bézier curves. It is implicit in what was said in the last section that the control points P_i determine a cubic path from P_0 to P_1

$$B(t) = (1 - t)^3 P_0 + 3t(1 - t)^2 P_1 + 3t^2(1 - t)P_2 + t^3 P_3$$

and that this path is a parametrization of the Bézier curve with these control points. In other words, $B(t)$ is a weighted combination of the control points. It is actually an average, which is to say that the sum of the coefficients is 1,

$$(1 - t)^3 + 3t(1 - t)^2 + 3t^2(1 - t) + t^3 = \big((1 - t) + t\big)^3 = 1,$$

by the binomial theorem for $n = 3$, which asserts that

$$(a + b)^3 = a^3 + 3a^2 b + 3ab^2 + b^3.$$

Because all the coefficients in our expression are nonnegative for $0 \le t \le 1$, $B(t)$ will lie inside the quadrilateral wrapped by the control points.

This idea will now be explored in more detail.

If $P_0, P_1, \ldots, P_{n-1}$ is a collection of n points in the plane, then a sum

$$c_0 P_0 + c_1 P_1 + \cdots + c_{n-1} P_{n-1}$$

is called a weighted average of the collection if (1) all the $c_i \geq 0$; (2) the sum of all the c_i is equal to 1. In the rest of this section, our primary goal will be to describe geometrically the set of all points we get as the weighted averages of a collection of points as the coefficients vary over all possibilities.

If $n = 2$, we know already that the set of all weighted averages $c_0 P_0 + c_1 P_1$ is the same as the line segment between P_0 and P_1 since we can write $c_1 = t$, $c_0 = (1 - t)$.

Suppose $n = 3$ and consider the weighted average

$$c_0 P_0 + c_1 P_1 + c_2 P_2.$$

Let's look at an explicit example. Look more precisely at

$$P = (1/4)P_0 + (1/4)P_1 + (1/2)P_2.$$

The trick we need to carry out is to rewrite this as

$$P = (1/2)[(1/2)P_0 + (1/2)P_1] + (1/2)P_2 = (1/2)Q + (1/2)P_2,$$

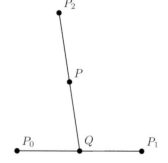

where $Q = (1/2)P_0 + (1/2)P_1$. In other words, P is the weighted average of the two points Q and P_2. The point Q is the weighted average of the original points P_0 and P_1 and hence must lie on the line segment between P_0 and P_1. In other words, we have the picture at the right.

Now we can almost always perform this trick, for we can write

$$c_0 P_0 + c_1 P_1 + c_2 P_2 = (c_0 + c_1)\left(\left(\frac{c_0}{c_0 + c_1}\right) P_0 + \left(\frac{c_1}{c_0 + c_1}\right) P_1\right) + c_2 P_2$$

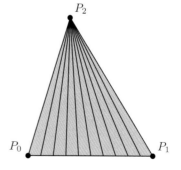

unless $c_0 + c_1 = 1 - c_2 = 0$. In the exceptional case we are just looking at P_2 itself, and in all other cases each weighted average of the three points is a weighted average of P_2 with a point on the line segment between P_0 and P_1. In other words, the set of all weighted averages of the three points coincides with the triangle spanned by the three points.

If we now look at four points, we get all the points on line segments connecting P_3 to a point in the triangle spanned by the first three. And in general we get all the points in a shape called the **convex hull** of the collection of points, which may be described very roughly as the set of points that would be contained inside a rubber band stretched around the whole collection and allowed to snap to them. The convex hull of a set of points in the plane or in space is very commonly used in mathematical applications, and plays a major role in computational graphics as well.

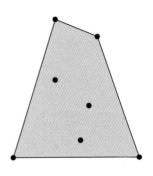

EXERCISE 6.9. *Write the simplest procedure you can with these properties: (1) it has two arguments x_0 and x_1 and (2) it draws the graph of $y = x^2$ between x_0 and x_1 with a single Bézier curve.*

EXERCISE 6.10. *Draw $y = x^5$ between $x = -1$ and $x = 1$ in the same way we drew $y = x^4$ earlier.*

6.9 HOW THE COMPUTER DRAWS BÉZIER CURVES

In this section we will see how the computer goes about drawing a Bézier curve. It turns out to be an extremely efficient process. First of all, a computer "thinks of" any path as a succession of small points (pixels) on the particular device it is dealing with. This is somewhat easier to see on a computer screen, certainly if you use a magnifying glass, but remains true even of the highest resolution printers. Thus, to draw something a computer just has to decide which pixels to color. It does this by an elegant recursive procedure – something akin to the following way to draw a straight line segment: (1) Color the pixels at each end. (2) Color the pixel at the middle. (3) This divides the segment into two halves. Apply steps (2) and (3) again to each of the halves and repeat them until the segments you are looking at are so small that they cannot be distinguished from individual pixels.

The analogous construction for Bézier curves, attributed to the car designer de Casteljau, goes like this:

Start with a Bézier curve with control points P_0, P_1, P_2, P_3. Perform the following construction. Set

$$P_{01} = \text{the median between } P_0 \text{ and } P_1$$
$$P_{12} = \text{the median between } P_1 \text{ and } P_2$$
$$P_{23} = \text{the median between } P_2 \text{ and } P_3$$

$$P_{012} = \text{the median between } P_{01} \text{ and } P_{12}$$
$$P_{123} = \text{the median between } P_{12} \text{ and } P_{23}$$
$$P_{0123} = \text{the median between } P_{012} \text{ and } P_{123}$$

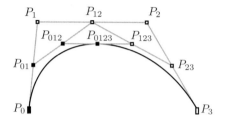 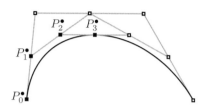

Then set

$$P_0^\bullet = P_0$$
$$P_1^\bullet = P_{01}$$
$$P_2^\bullet = P_{012}$$
$$P_3^\bullet = P_0^{\bullet\bullet} = P_{0123}$$
$$P_1^{\bullet\bullet} = P_{123}$$
$$P_2^{\bullet\bullet} = P_{23}$$
$$P_3^{\bullet\bullet} = P_3.$$

The point $P_3^\bullet$ turns out to lie on the Bézier curve determined by the original points P_i at approximately the halfway point. The Bézier curve can now be split into halves, each of which is itself a Bézier cubic, and the control points of the two new curves are among those constructed above. The two halves might be called the Bézier curves **derived** from the original. The points $P_i^\bullet$ are those for the first half, the $P_i^{\bullet\bullet}$ for the second. If we keep subdividing in this way we get a sequence of midpoints for the smaller segments (actually a kind of branched list), and to draw the curve we just plot these points after the curve has been subdivided far enough. This sort of subdivision can be done very rapidly by a computer since dividing by two is a one-step operation in base 2 calculations, and in fact drawing the pixels to go on a straight line is not much faster.

EXERCISE 6.11. *Draw the figure above with PostScript.*

EXERCISE 6.12. *The point $P_1^\bullet$ is $(1/2)P_0 + (1/2)P_1$. Find similar expressions for all the points constructed in terms of the original four.*

EXERCISE 6.13. *The purpose of this exercise is to prove that each half of a Bézier curve is also a Bézier curve. Let*

$$P(s) = (1 - s)^3 P_0 + 3s(1 - s)^2 P_1 + 3s^2(1 - s)P_2 + s^3 P_3.$$

The point is to verify that this formula agrees with the geometrical process described above. Let $P_1^\bullet$, and so on, be the points defined just above. The first half of the Bézier curve we started with is a cubic curve with initial parameter 0 and final parameter $1/2$. Let $\Delta t = 1/2$. Verify that

$$P(0) = P_0 \quad \text{(trivial)}$$
$$P'(0) = (\Delta t/3)(P_1^\bullet - P_0) \quad \text{(almost trivial)}$$
$$P(1/2) = P_3^\bullet$$
$$P'(1/2) = (\Delta t/3)(P_3^\bullet - P_2^\bullet).$$

These equations, by the earlier characterization of control points in terms of derivatives, guarantee that the first half of the original Bézier path is a Bézier path with control points $P_0 = P_0^\bullet$, $P_1^\bullet$, $P_2^\bullet$, $P_3^\bullet$.

EXERCISE 6.14. *How might a computer construct quadratic Bézier curves in a similar way?*

6.10   BERNSTEIN POLYNOMIALS

The Bézier cubic polynomial

$$y_0(1 - t)^3 + 3y_1(1 - t)^2 t + 3y_2(1 - t)t^2 + y_3 t^3$$

is just a special case of a more general construction of **Bernstein polynomials**. In degree one we have the linear interpolating function

$$y_0(1 - t) + y_1 t,$$

in degree two we have the quadratic functions mentioned earlier, and in degree n we have the polynomial

$$B_y(t) = y_0(1 - t)^n + ny_1 t(1 - t)^{n-1} + \frac{n(n - 1)}{2} y_2 (1 - t)^{n-2} t^2 + \cdots + y_n t^n,$$

where y is the array of the **control values** y_i and the other coefficients make up the nth row of Pascal's triangle

$$
\begin{array}{ccccccc}
 & & & 1 & & & \\
 & & 1 & & 1 & & \\
 & 1 & & 2 & & 1 & \\
1 & & 3 & & 3 & & 1 \\
 & & & \cdots & & &
\end{array}
$$

These polynomials were first defined by the Russian mathematician Sergei Bernstein in the early twentieth century to answer a sophisticated question in approximation theory.

These also are weighted sums of the control values, and so for $0 \leq t \leq 1$ the value of $B_y(t)$ will lie in the range spanned by the y_i. In particular, if the y_i are a nondecreasing sequence

$$y_0 \leq y_1 \leq \cdots y_n,$$

then $y_0 \leq B_y(t) \leq y_n$ for $0 \leq t \leq 1$. But much more can be said.

EXERCISE 6.15. *Prove that*

$$B'_y(t) = n B_{\Delta y}(t),$$

where Δy is the array of differences

$$\Delta y = (y_1 - y_0, y_2 - y_1, \ldots, y_n - y_{n-1}).$$

EXERCISE 6.16. *Prove that, if the y_i are nondecreasing then $B_y(t)$ is a nondecreasing function over the range $[0, 1]$.*

EXERCISE 6.17. *There is a way to evaluate $B_y(t)$ for $0 \leq t \leq 1$ along the lines used by the computer to construct the Bézier cubic curve. It can be described best in a recursive fashion. First of all, if y has length 1, the Bernstein polynomial is just a constant. Otherwise, with $n > 0$, form a derived sequence of length $n - 1$:*

$$\delta y = ((1 - t)y_0 + ty_1, \ldots, (1 - t)y_{n-1} + ty_n).$$

Then

$$B_y(t) = B_{\delta y}(t).$$

Prove this. Explain how this process is related to the naïve construction of Pascal's triangle, one row at a time.

6.11 THIS SECTION BRINGS YOU THE LETTER O

Paths can be constructed in PostScript in various ways through commands such as moveto, but internally PostScript stores a path as an array, storing exactly four different types of objects – moveto, lineto, curveto, closepath tags together with the arguments of the command. This array can be accessed explicitly by means of the command pathforall. This command has four arguments, each of which is a procedure. It loops through all the components of the current path, pushing appropriate data on the stack and then applying the procedures respectively to moveto, lineto, curveto, and closepath components. For moveto and lineto components it pushes the corresponding values of x and y in current user coordinates; for curveto it pushes the six values of x_1, y_1, and so on (also in user coordinates); and for closepath it pushes nothing. The following procedure, for example, displays the current path:

```
/path-display {
  { [ 3 1 roll (moveto) ] == }
  { [ 3 1 roll (lineto) ] ==  }
  { [ 7 1 roll (curveto) ] == }
  { [ (closepath) ] == }
  pathforall
} def
```

The following procedure tells whether a current path has already been started since it returns with true on the stack if and only if the current path has at least one component:

```
/thereisacurrentpoint{
  false {
    { 3 { pop } repeat true exit }
    { 3 { pop } repeat true exit }
    { 7 { pop } repeat true exit }
    { pop true exit }
  } pathforall
} def
```

The most interesting paths in PostScript are probably strings – that is, the paths formed by strings when the show operator is applied or, in other words, the path the string will make when it is drawn in the current font. This outline can be accessed as a path by applying the command charpath, which has two arguments. The first is a string. The second is a Boolean variable, which is more or less irrelevant to our purposes. The command appends the path described by the string in the current graphics environment to the current path if it is assumed, in particular, that a font has been selected. In this way, for example, you can deal with the outlines of strings as if they were ordinary paths. The code

```
/Times-Roman findfont
40 scalefont
setfont

newpath
0 0 moveto
(Times-Roman) false charpath
gsave
1 0 0 setrgbcolor
fill
grestore
stroke
```

produces

Times-Roman

You can combine `charpath` and `pathforall` to see the explicit path determined by a string, but only under suitable conditions. Many, if not most, PostScript fonts have a security mechanism built into them that does not allow the paths of their characters to be deconstructed, and you will get an error from `pathforall` if you attempt to do so. So if you want to poke around in character paths you must be working with a font that has not been declared inaccessible. This is not a serious restriction for most of us since there are many fonts, including the ones usually stocked with GhostScript, that are readable. Here is the path of the character "O" from the font called `/Times-Roman` by GhostScript:

```
0.360998541 0.673999 moveto
0.169997558 0.673999 0.039997559 0.530835 0.039997559 0.33099854 curveto
0.039997559 0.236999512 0.0697631836 0.145998538 0.119995117 0.0879980475 curveto
0.177995607 0.0249975584 0.266994625 -0.0140014645 0.354995131 -0.0140014645 curveto
0.551994622 -0.0140014645 0.689997554 0.125998542 0.689997554 0.326999515 curveto
0.689997554 0.425998539 0.660305202 0.510998547 0.603996575 0.570998549 curveto
0.540996075 0.639997542 0.456999511 0.673999 0.360998541 0.673999 curveto
closepath
0.360996097 0.63399905 moveto
0.406997085 0.63399905 0.452998042 0.618159175 0.488999 0.58999753 curveto
0.542998075 0.540998518 0.58 0.44699952 0.58 0.328000486 curveto
0.58 0.269001454 0.564633787 0.200002447 0.540998518 0.148002923 curveto
0.531999528 0.123002931 0.515 0.098002933 0.491999507 0.0750024393 curveto
0.456999511 0.0400024429 0.411999524 0.0260009766 0.358999 0.0260009766 curveto
```

```
0.312998056 0.0260009766 0.26799804 0.04253418 0.232998043 0.0710034147 curveto
0.180998534 0.117004395 0.15 0.218005374 0.15 0.329006344 curveto
0.15 0.431005865 0.177197263 0.528005362 0.217998043 0.575004876 curveto
0.256997079 0.618005395 0.30599609 0.63399905 0.360996097 0.63399905 curveto
closepath
0.721999526 -0.0 moveto
```

and here is the first Bézier curve in the path:

REFERENCES

1. R. E. Barnhill and R. F. Riesenfeld (editors), **Computer Aided Geometric Design**, Academic Press, 1974. This book contains papers presented at a conference at the University of Utah that initiated much of modern computer graphics. The article by P. Bézier is very readable.

2. G. Farin, **Curves and Surfaces for Computer Aided Design**, Academic Press, 1988. This is a pleasant book that probably covers more about curves and surfaces than most readers of this book will want, but the first chapter is an enjoyable account by P. Bézier on the origins of his development of the curves that bear his name. These curves were actually discovered much earlier (before computers were even a dream) by the mathematicians Hermite and Bernstein, but it was only the work of Bézier, who worked at the automobile maker Renault, and de Casteljau, who worked at Citroen, that made these curves familiar to graphics specialists.

3. D. E. Knuth, METAFONT: **The Program**, Addison-Wesley, 1986. Pages 123–131 explain extremely clearly the author's implementation of Bézier curves in his program METAFONT. For the admittedly rare programmer who wishes to build his or her own implementation (at the level of pixels), or for anyone who wants to see what attention to detail in first-class work really amounts to, this is the best resource available.

Interlude

At this point you have seen just about all the basic PostScript commands you'll ever use. The rest of this book will be spent showing you how to combine them to make some very complicated – one would hope even very beautiful – figures.

You should by now have no trouble drawing simple figures but are probably starting to worry about how to do more difficult things. I want to offer you some advice – in part just to collect in one place remarks I have made throughout the text. Much of what I have to say is not specific to PostScript but might well be made about programming in general.

- Most programs you write will be just an assembly of smaller chunks that are themselves quite simple. Your coding should reflect this, making the chunks in your code as independent of each other as possible. Just as your pages should be independent of each other.

- The reason you should try to arrange your program so it looks as much as possible like an assembly of smaller chunks is that you can then concentrate on getting each chunk completely correct. The most important thing to keep in mind in good programming is, as the real estate agents (don't) say, *locality*, *Locality*, *LOCALITY*. The effects of your code should be carefully set up to be local, affecting if possible only data needed at the moment it is running. One example of this I have already remarked on is that procedures, above all, which may be called from anywhere in your code, should use local variables and have few or no side effects. Those it does have should be clearly specified.

- Procedures should be as isolated as possible from the rest of your code. It is best to put them in separate files and run these. Then you can test your procedures independently of the rest of your program. With the PERL script described in Appendix 2, embedding files that are run during development into the final program is simple.

- In particular, make up a file defining your favourite constants like e, π and so on and run that file to obtain access to them. One thing I haven't mentioned is that for efficiency you can get PostScript to embed these numbers directly if you write using //. For example, //pi will immediately substitute 3.1415... if you have previously written /pi 3.1415926535 def. (Whereas ordinarily it defers evaluating the expression pi because you might very well have redefined it, and doesn't know that you mean it to be a constant.) I don't suppose, for the programs we are concerned with here, that this really increases speed much, but it makes you feel good.

- As a programming language, PostScript is special because of its direct link to graphics. Use this feature. When starting to draw a picture, begin by getting *something* up on the screen that looks roughly like what you want and then begin to modify it. Visual debugging compensates somewhat for the otherwise terrible debugging environment of PostSCript.

- Debugging PostScript using Ghostscript is nasty. The only way to avoid it, however, is to write only perfect lines of code that never need rewriting. But for those presumably rare moments when things aren't going quite right, you'll have to descend to the land of mortals. So far I have mentioned the techniques of spilling out data in the terminal window and running gsnd. To make this easier I myself use a procedure display with one argument n that spills out in an array, without destruction, the top n items on the stack:

```
/display { 1 dict begin
  /n exch def
  n copy [ n 1 add 1 roll ] ==
end } def
```

I put this in a file display.inc and run it at the top of nearly all my programs. For line breaks to make output more readable, use () =. You can also use quit to break your code off at a selected spot.

- Keep your stack clean. A common error is to forget to take everything off the stack in procedures. You can check this by running gs, which indicates the stack size at the end.

- Remember that coordinate changes are cumulative.

- The part of your code that actually does the drawing should be as clean and readable as possible. Do all necessary calculations ahead of time. Path drawing is the cockpit of PostScript programming. Leave unnecessary items at the door when you enter.

- At the beginning of a project, use lots of variables and procedures. Readability at that stage is extremely important. Comment freely. Slim down your code when and if necessary.

- Make your code readable, not only by adding comments but by separating different parts by dividers such as

```
% --- this part does blah blah ---------------------------
```

so you can scan your file easily to get where you want to go. As for comments, the most important ones are those that describe procedures – tell what arguments they need, what they return, and what side effects they have. Procedures will usually be called many times in many different environments, and you will not likely want to read the whole procedure over again to figure out what it's doing.

CHAPTER 7

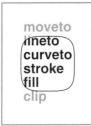

Drawing curves automatically: procedures as arguments

The process of drawing curves by programming each one specially is too complicated to be done easily. In this chapter we will see how to construct procedures that help out enormously. We proceed in stages, starting with a reasonably simple example.

7.1   DRAWING AN HYPERBOLA

The curves we have drawn so far are really too simple to give you an idea of how useful Bézier curves are. This is because the Bézier curve is guaranteed to match exactly the graph of any polynomial of degree 3 or less, and so no approximation is involved. In this section we will develop a procedure `hyperbola` with three arguments – x_0, x_1, N – that constructs the graph of the upper branch of an hyperbola

$$y^2 - x^2 = 1, \quad y = \sqrt{1 + x^2}$$

by using N Bézier segments in between x_0 and x_1.

As is usually the best idea, the procedure builds the path without drawing it. Thus, the sequence

```
newpath
-2 2 4 hyperbola
stroke
```

will draw the curve $y = \sqrt{1 + x^2}$ from $x = -2$ to $x = 2$ in four Bézier segments.

Paths drawn by −2 2 1 hyperbola *(pink) and* −2 2 2 hyperbola *(red).*

What goes into the procedure hyperbola? We can immediately write down the skeleton

```
/hyperbola { 16 dict begin
   /N exch def
   /x1 exch def
   /x0 exch def
   ...
end } def
```

and we must now fill in the real computation. First we set a step size $h = (x_1 - x_0)/N$ so that in N steps we cross from x_0 to x_1:

```
/h x1 x0 sub N div def
```

Then we introduce variables x and y, which are going to change as the procedure runs, and move to the first point on the graph. It will also help to keep a variable s to hold the current value of the slope. Note that if

$$y = f(x) = \sqrt{1 + x^2} = (1 + x^2)^{1/2},$$

then

$$s = f'(x) = (1/2)(2x)(1 + x^2)^{-1/2} = \frac{x}{\sqrt{1 + x^2}} = \frac{x}{y}.$$

Recall that the control points are

$$(x_0, y_0)$$
$$(x_0 + h/3, y_0 + y'(x_0)/3$$
$$(x_1 - h/3, y_1 - y'(x_1)/3$$
$$(x_1, y_1)$$

where $x_1 = x + h$, $y_1 = y(x_1)$. The code begins as follows:

```
/x x0 def
/y 1 x x mul add sqrt def
/s x y div def
x y moveto
```

Now we must build N Bézier segments, using a `repeat` loop

```
N { % repeat
  x h 3 div add
  y h 3 div s mul add
  /x x h add def
  /y 1 x x mul add sqrt def
  /s x y div def
  x h 3 div sub
  y h 3 div s mul sub
  x y
  curveto
} repeat
```

and that's it.

We could make this program somewhat more readable and more flexible if we added a few procedures to it that calculate $f(x)$ and $f'(x)$ instead of doing it in line. Each of the procedures should have a single argument x. They are short enough that we do not need to use a variable inside them. Explicitly,

```
 % sqrt(1 + x^2)
/f {
  dup mul 1 add sqrt
} def
 % x/sqrt(1 + x^2)
/f' {
  dup dup mul 1 add sqrt div
} def
```

Recall that dup just duplicates the item on the top of the stack. The new loop would
be

```
N { % repeat
  x h 3 div add
  y h 3 div s mul add
  /x x h add def
  /y x f def
  /s x f' def
  x h 3 div sub
  y h 3 div s mul sub
  x y
  curveto
} repeat
```

It would be better to have a single procedure that calculates $f(x)$ and $f'(x)$ all
in one package. For one thing, it would be more efficient since we wouldn't have
to calculate square roots more often than necessary. For another, I prefer to have
things that go together . . . well, go together. The calculations of $f(x)$ and $f'(x)$
are related, and if you modify one to draw some different graph, then you will have
to modify the other. For this reason, they should be dealt with in one visible and
indivisible unit. We can do this by using a single procedure with one argument
x and as output an array of two numbers $[y \; s]$. But now a variable to hold the
value of x is useful. It might be a good idea here to exhibit all procedures we are
using:

```
% x -> [sqrt(1 + x^2), x/sqrt(1 + x^2)]
/f { 2 dict begin
  /x exch def
  /y 1 x x mul add sqrt def
[
  y
  x y div
]
end } def

% x0 x1 N
/hyperbola { 16 dict begin
  /N exch def
  /x1 exch def
```

```
/x0 exch def
 % h = (x1 - x0)/N
/h x1 x0 sub N div def
/x x0 def
/F x f def
/y F 0 get def
/s F 1 get def
x y moveto
N { % repeat
  x h 3 div add
  y h 3 div s mul add
  /x x h add def
  /F x f def
  /y F 0 get def
  /s F 1 get def
  x h 3 div sub
  y h 3 div s mul sub
  x y
  curveto
 } repeat
end } def
```

It is true that using a dictionary and an array has made the program somewhat less efficient than it was at the start. On the good side, the program is now perhaps slightly more readable – or perhaps not, as determined by your own taste. It has one great virtue, however: it is a great deal more flexible. *If we want to draw some other graph, we need only rewrite the single procedure* f, *making sure that it, too, has a single argument x and returns an array of two values* [y s].

EXERCISE 7.1. *Modify the procedure* f *so you can use essentially the same stuff to graph the function* $y = x^3 - x$ *between* -1 *and* 1.

EXERCISE 7.2. *Modify the procedure* f *so you can use essentially the same stuff to graph the function* $y = \sin x$ *between* -1 *and* 1. *Be careful about degrees and radians; it is only when x is expressed in radians that* $\sin' x = \cos x$. *Use the modified procedure to draw the graph of* $\sin x$ *between* $x = 0$ *and* $x = \pi$ *with* 1, 2, 4 *segments, all on one plot. Make them successively lighter so you can distinguish them.*

7.2   PARAMETRIZED CURVES

We now have a good idea of how to draw smooth function graphs. However, not all curves in the plane are the graphs of functions. What is true is that almost any curve we can imagine is the union of several smooth segments, where each segment is a parametrized path $(x(t), y(t))$ in the plane.

I recall that to draw a segment between $t = t_0$ and t_1 by Bézier curves requires control points

$$
\begin{aligned}
P_0 &= (x_0, y_0) \\
&= (x(t_0), y(t_0)) \\
P_1 &= (x_1, y_1) \\
&= (x(t_1), y(t_1)) \\
P_{1/3} &= \begin{bmatrix} x_{1/3} \\ y_{1/3} \end{bmatrix} \\
&= \begin{bmatrix} x_0 \\ y_0 \end{bmatrix} + \frac{\Delta t}{3} \begin{bmatrix} x_0' \\ y_0' \end{bmatrix} \\
P_{2/3} &= \begin{bmatrix} x_{2/3} \\ y_{2/3} \end{bmatrix} \\
&= \begin{bmatrix} x_1 \\ y_1 \end{bmatrix} - \frac{\Delta t}{3} \begin{bmatrix} x_1' \\ y_1' \end{bmatrix}.
\end{aligned}
$$

Let's look again at the problem posed at the beginning of this section and see how we would draw a parametrized path by using Bézier curves. We must first divide it up into a certain number of segments. Calculate the position and velocity at the endpoints of each of the segments. Move to the first point and add one Bézier curve for each segment. (The command moveto is not necessary in the intermediate curves because drawing a Bézier curve in PostScript advances the current point to the endpoint.) Then stroke it, fill it, or clip it.

7.3   DRAWING GRAPHS AUTOMATICALLY

In the next section I'll explain a technique for drawing parametrized curves that will become a standard trick in your bag of tools. There are several new ingredients in it, and it may help if I explain one of them by improving the procedure for drawing hyperbolas along the same lines. We have already made that procedure reasonably flexible by isolating how the actual function $f(x)$ is used to draw the graph. What I will do now is show how *you can change functions by using the graphing function itself as an argument to a general procedure for making graphs.* The *only* part of the procedure hyperbola that must be changed is the very first part. Since the procedure no longer draws only hyperbolas, its name must be changed. And it has an extra argument, the **name** of the function f, which must be a procedure into which you put x and out of which you get an array $[f(x)\ f'(x)]$. We must read this fourth argument and convert that name into a procedure so we can call it. The few lines of this procedure where it differs from the older one are as follows:

```
/mkgraph {
  load
  16 dict begin
  /f exch def
  /N exch def
  /x1 exch def
  /x0 exch def
```

The process is not quite as simple as might have been expected. A slight technical difficulty arises because the name of the parametrization procedure being used here might be redefined inside the procedure – it might, for example, be /f – and so it must be retrieved from its dictionary with load before a new dictionary is introduced. Here is a complete program that uses this to draw the graph of $y = x^4$.

```
% x0 x1 N /f
/mkgraph { load
  /f exch def
  1 dict begin
  /N exch def
  /x1 exch def
  /x0 exch def
   % h = (x1 - x0)/N
  /h x1 x0 sub N div def
  /x x0 def
  /F x f def
  /y F 0 get def
  /s F 1 get def
  x y moveto
  N {
    x h 3 div add
    y h 3 div s mul add
    /x x h add def
    /F x f def
```

```
        /y F 0 get def
        /s F 1 get def
        x h 3 div sub
        y h 3 div s mul sub
        x y
        curveto
      } repeat
    end } def


    % [x^4 4x^3]
    /quartic { 2 dict begin
      /x exch def
      [
        x x mul x mul x mul
        x x mul x mul 4 mul
      ]
    end } def


    % -------------------------------------------


    72 72 scale
    4.25 5.5 translate
    1 72 div setlinewidth

    newpath
    -1 1 8 /quartic mkgraph
    stroke
```

In the next section we'll see a procedure that is rather similar to this one but differs in these aspects: (1) It deals with parametrized curves instead of graphs; (2) it allows you to use a single procedure to draw any one of a large family of curves, such as all of the graphs $y = cx^4$, where c is a constant you can specify when you draw the curve; (3) it adds the new path to the path that already exists, if there is one, like the command arc.

EXERCISE 7.3. *Write a PostScript procedure with the same arguments as* mkgraph *but that simply draws a polygon among the successive points. (This can be used to debug your calculus.)*

EXERCISE 7.4. *Write a PostScript procedure that will graph a polynomial between x_0 and x_1 with N Bézier segments. There are a number of things you have to think about: (1) For evaluating a polynomial*

*in a program it is easiest to use an expression like $5x^3 + 2x + 3x + 4 = ((5x + 2)x + 3)x + 4$.
(2) You will have to add an argument to this procedure to pass the polynomial coefficients as an
array. Recall that* length *returns the size of an array. (See Appendix 6 for more about polynomial
evaluation.)*

7.4   DRAWING PARAMETRIZED PATHS AUTOMATICALLY

If you are given a parametrized path and you want to draw it by using Bézier curves,
you must calculate position and velocity at several points of the path. This is tedious
and prone to error, and you will quickly ask if there is some way to get PostScript
to do the work. This is certainly possible if you can write a PostScript routine that
calculates position and velocity for the parametrization. One tricky point is that we
don't want to rewrite the drawing routine for every path we draw but would like
instead to put in the **parametrization** itself as an argument passed to the routine.
The parametrization should be a procedure that has a single argument, namely
a value of t, and returns data giving position and velocity for that value of t. We
will in fact do this and make the name of the routine that calculates position and
velocity one of the arguments. How PostScript handles this is somewhat technical,
but you won't have to understand underlying details to grasp the routine. Another
tricky point is that we would like to make a routine for drawing a family of curves;
we don't want to have to make separate routines for $y = x^2$, $y = 2x^2$, and so on.
We would like to be able at least to write one routine that can draw $y = cx^2$ for any
specified constant c. This will be accomplished by passing an array of **parameters**
to the routine to pick a specific curve from a family. The terminology is a bit clumsy,
for we have the parameters determining which path from a family is to be drawn
and the variable t that parametrizes the curve. I will call the last, to avoid confusion,
the **parametrizing variable**.

I will first simply lay out the main routine to be used from now on for drawing
parametrized paths. It is somewhat complicated. In the next two sections I will
explain how to use it and how it works.

```
% stack: t0 t1 N [parameters] /f
/mkpath { load
1 dict begin
/f exch def
/pars exch def
/N exch def
/t1 exch def
/t0 exch def
    % h = (t1-t0)/N
```

```
  /h t1 t0 sub N div def
   % h3 = h/3
  /h3 h 3 div def
   % set current location = [f(t0) f'(t0)]
  /currentloc pars t0 f def
  pars t0 f 0 get
  aload pop     % calculate the first point
  thereisacurrentpoint     % if a path already under construction ...
    { lineto }
    { moveto }

  ifelse
  N {                                % x y = currentpoint
    currentloc 0 get 0 get            % x0 dx0
    currentloc 1 get 0 get
    h3 mul
    add
     % x1
    currentloc 0 get 1 get
    currentloc 1 get 1 get
    h3 mul
    add
     % y1
    /t0 t0 h add def
     % move ahead one step
    /currentloc pars t0 f def
    currentloc 0 get 0 get
    currentloc 1 get 0 get
    h3 mul sub
    currentloc 0 get 1 get
    currentloc 1 get 1 get
    h3 mul sub
     % x2 y2
    currentloc 0 get 0 get
    currentloc 0 get 1 get
     % x3 y3
    curveto
  } repeat
end } def
```

The procedure thereisacurrentpoint returns true or false, depending on whether a path has been started or not. We have already seen it in Chapter 6.

HOW TO USE IT

The input to the procedure consists of five items on the stack.

- *The first is the initial value of t for the path.*
- *The second is the final value of t.*
- *The third is the number N of Bézier segments to be drawn.*
- *Fourth comes an array [...] of parameters, which I will say something more about in a moment. It can be just the empty array [], but it must be there.*
- *Last follows the name of a routine specifying the parametrization in PostScript. A name in PostScript starts with the symbol /. As I have already said, this routine has two arguments. The first is an array of things that the routine can use to do its calculation. The second is the variable t. Its output (left on the stack) will be a 2 × 2 matrix written according to my conventions as an array of two 2D arrays.*

The most important, and complicated, item here is the parametrization routine.

For example, suppose we want to draw circles of varying radii centered at the origin. The parametrization of such a circle is

$$t \mapsto P(t) = (R \cos t, \ R \sin t),$$

and the velocity vector of this parametrization is

$$t \mapsto P'(t) = (-R \sin t, \ R \cos t)$$

if t is in radians. The variable t is the parameter that traverses the path, whereas R is a parameter specifying which of several possible circles are to be drawn. Thus, the input to the circle-drawing routine will be a pair [R] t and the output will be [[xt yt][dxt dyt]].

Here is a more explicit block of PostScript code (on the assumption that pi was defined elsewhere to be π). Note that PostScript uses degrees instead of radians but that the formula for the velocity vector assumes radians.

```
/circle { 4 dict begin
  /t exch def
  /pars exch def
  /R pars 0 get def
  /t t 180 mul //pi div def
  [
    [ t cos R mul t sin R mul ]
```

```
        [ t sin neg R mul t cos R mul ]
    ]
end } def

(mkpath.inc) run

newpath
[2] /circle 0 2 //pi mul 8 mkpath
closepath
stroke
```

You might not have seen anything like //pi before; using // preceding a variable name means the interpreter puts its current value in place immediately rather than waiting until the procedure is run to look up pi and evaluate it.

The array of parameters passed to one of these routines can be of any size you want. It can even be empty if in fact you just want to draw one of a kind. But you will probably find that most paths you want to draw are just part of a larger family.

There is one other point. It is very common to get the routine for the velocity vector wrong because, after all, you must first calculate a derivative. When this happens, the curve will hit the points representing position but will wander wildly in between. One way to see if in fact you have computed the derivative correctly is to use a routine that might be called mkpolypath, which simply draws a polygonal path instead of one made up of Bézier segments. This routine has exactly the same usage as mkpath but ignores the velocity vector in building the path and can hence be used to see if the rest of your routine is working.

EXERCISE 7.5. *Write a procedure* mkpolypath *that has the same arguments as* mkpath *but draws a polygon instead.*

EXERCISE 7.6. *Write down a parametrization of the ellipse*

$$\frac{x^2}{a^2} + \frac{y^2}{b^2} = 1.$$

Write a procedure that will draw it. For example you would write 3 4 drawellipse.

EXERCISE 7.7. *Draw the image of the* 12×12 *grid centered at* $(0, 0)$ *under the map* $(x, y) \mapsto (x^2 - y^2, 2xy)$. *The spacing between grid lines is to be* 0.25 *cm.*

7.6   HOW IT WORKS

The basic idea is simply to automate the procedure you used earlier to draw the graph of $y = x^4$. The point is that if we are given a parametrization of a path we can draw an approximation by Bézier curves using the velocity vectors associated to the parametrization to construct control points. The routine is fairly straightforward except that it calls a procedure `thereisacurrentpoint` to tell whether the path being drawn is the beginning of a new path or the contin-  uation of an old one. You don't have to know the details of the procedure called; it operates in a very simple manner at a somewhat low level of PostScript.

This routine all by itself is very useful and is capable of making interesting pictures. But the ideas behind it are applicable in a wide range of circumstances.

7.7   CODE

The file `mkpath.inc` contains a procedure `mkpath` as well as `mkgraph` and a few other related ones.

CHAPTER 8

Nonlinear 2D transformations: deconstructing paths

Sometimes we want to draw a figure after a nonlinear transformation has been applied to it. In image manipulation programs, this is often called **morphing**. For example, here is a morphed 10×10 grid produced by a program in which the basic drawing commands drew a square grid, and these were followed by some transforming code before stroking.

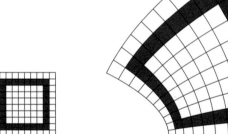

To apply transformations to paths, we just have to understand (a) transformations and (b) paths!

8.1 TWO-DIMENSIONAL TRANSFORMATIONS

A 2D transformation is a function $f(x, y)$ of two variables that returns a pair of numbers $u(x, y)$ and $v(x, y)$, the coordinates of the transform of the point (x, y). We have already seen affine transformations where

$$f(x, y) = (ax + by + c, dx + ey + f)$$

for suitable constants a, b, and so on. But now we want to allow more complicated ones. I should say right at the beginning that these can be very complicated. An affine transformation is not so difficult to visualize because we know what the transformation does everywhere if we know what it does to just a single square. But an arbitrary transformation may have very different effects in different parts of the plane, and this is the source of much difficulty in comprehending it. Indeed, the nature of 2D transformations has been in not-so-distant times the subject of interesting mathematical research. (I am referring to the stability of properties of such transformations under perturbation, part of the so-called catastrophe theory.)

I'll spend some time looking at one that is not too complicated:

$$f(x, y) = (x^2 - y^2, 2xy).$$

This is not quite a random choice; it is derived from the function of complex numbers that takes z to z^2 because

$$(x + iy)^2 = (x^2 - y^2) + i\,(2xy).$$

The best way to understand what it does is to write (x, y) in polar coordinates as $(r\cos\theta, r\sin\theta)$ since in these terms f takes (x, y) to

$$(r^2\cos^2\theta - r^2\sin\theta, 2r^2\cos\theta\sin\theta) = (r^2\cos 2\theta, r^2\sin 2\theta).$$

In other words, it squares r and doubles θ. We can see how this works in the following figures. On the left is the sector of the unit circle between 0 and $\pi/2$, on the right its image with respect to f:

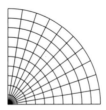
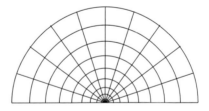

That seems simple enough. And yet, the image that began this chapter is the transformation with respect to this same f of the square with lower left corner at $(1/2, 0)$, upper right at $(3/2, 1)$, and sides aligned with the axes. So even this simple transformation has some interesting effects. We'll see others when we look at what it does to character strings.

A transformation is completely described by the formula for its separate components, that is, the functions u and v in the expression

$$f(x, y) = (u(x, y), v(x, y)).$$

This is all we'll need to know about f to apply it. But to understand in any deep sense how a transformation in 2D behaves, we also must consider its **Jacobian derivative**. This is a matrix-valued function of x and y whose entries are the partial derivatives of f:

$$\text{Jac}_f(x, y) = \begin{bmatrix} \dfrac{\partial u}{\partial x} & \dfrac{\partial v}{\partial x} \\ \dfrac{\partial u}{\partial y} & \dfrac{\partial v}{\partial y} \end{bmatrix}.$$

For the example we have already looked at this is

$$\begin{bmatrix} 2x & 2y \\ -2y & 2x \end{bmatrix}.$$

The importance of the Jacobian matrix is that it allows us to approximate the map f by an affine map in any small neighborhood of a point. The exact statement is

■ *If Δx and Δy are small, then we have an approximate equality*

$$f(x + \Delta x, y + \Delta y) \approx f(x, y) + [\,\Delta x \quad \Delta y\,]\,\text{Jac}_f(x, y).$$

The following is another way to say it:

■ *If (x, y) lies near (x_0, y_0), then we have an approximate equality*

$$f(x, y) \approx f(x_0, y_0) + [\,(x - x_0) \quad (y - y_0)\,]\,\text{Jac}_f(x, y).$$

What this says, roughly, is that locally in small regions the transformation looks as though it were affine.

This phenomenon can be seen geometrically. If we look at the figure that started this chapter, but zoom in, we see the following:

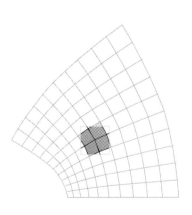
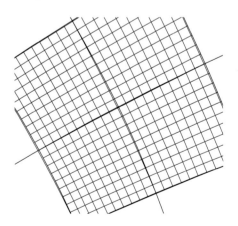

In other words, if we zoom in to a part of the plane, the transformation looks straighter and straighter as the scale shrinks. I'll not say much more about why this **Jacobian approximation** is valid but only remark that it is essentially the definition of partial derivatives. It is the generalization to 2D of the linear approximation $f(x + \Delta x) \approx f(x) + f'(x)\Delta x$ encountered in the ordinary calculus of one variable.

In our example, we have an exact equation

$$f(x + \Delta x, y + \Delta y) = ((x + \Delta x)^2 - (y + \Delta y)^2, 2(x + \Delta x)(y + \Delta y))$$
$$= ((x^2 + 2x\Delta x + \Delta x^2) - (y^2 + 2y\Delta y + \Delta y^2),$$
$$2(xy + x\Delta y + y\Delta x + \Delta x\Delta y),$$

which becomes

$$((x^2 + 2x\Delta x) - (y^2 + 2y\Delta y), 2(xy + x\Delta y + y\Delta x)$$
$$\approx (x^2 - y^2, 2xy) + [2x\Delta x - 2y\Delta y, 2x\Delta y + y\Delta x]$$
$$= f(x, y) + [\Delta x \quad \Delta y] \begin{bmatrix} 2x & 2y \\ -2y & 2x \end{bmatrix}$$

if we ignore the terms of order two – that is, Δx^2, Δy^2, $\Delta x\Delta y$. But this **linear approximation** is the Jacobian approximation. As it always is.

I'll not use the Jacobian derivative in PostScript procedures, although it could improve quality at the cost of complication. But it is important to realize that the reason our transformation procedures will work so well is precisely because in very small regions the function f looks affine.

In routines to be described later, a 2D transformation f will be given as a procedure with two arguments x and y, and it will return a pair u v. In our example it would be

```
% on stack: x y
/fcn { 2 dict begin
  /y exch def
  /x exch def
  x dup mul y dup mul sub % x^2-y^2
  2 x mul y mul % 2xy
end } def
```

It would in principle be possible to get a smoother output if I added a second version that takes advantage of the Jacobian approximation. In this version the function *f* would return the Jacobian matrix on the stack as well in the form of an array of four variables. In practice this doesn't seem to be important.

8.2 CONFORMAL TRANSFORMS

As we have seen, the Jacobian matrix of the transform $(x, y) \mapsto (x^2 - y^2, 2xy)$ is

$$\begin{bmatrix} 2x & 2y \\ -2y & 2x \end{bmatrix}.$$

This matrix has the form

$$\begin{bmatrix} a & b \\ -b & a \end{bmatrix},$$

and such matrices, which I'll call **similarity matrices** for reasons that will become apparent in a moment, have interesting properties. One example of such a matrix is the rotation matrix

$$\begin{bmatrix} \cos\theta & \sin\theta \\ -\sin\theta & \cos\theta \end{bmatrix},$$

and another is the scalar matrix

$$\begin{bmatrix} r & \\ & r \end{bmatrix},$$

So also is the product of these two, which is

$$\begin{bmatrix} r\cos\theta & r\sin\theta \\ -r\sin\theta & r\cos\theta \end{bmatrix}.$$

In fact, if *S* is any similarity matrix

$$\begin{bmatrix} a & b \\ -b & a \end{bmatrix},$$

we can write

$$\frac{a}{\sqrt{a^2 + b^2}} = \cos\theta$$

$$\frac{b}{\sqrt{a^2 + b^2}} = \sin\theta$$

and then

$$\begin{bmatrix} a & b \\ -b & a \end{bmatrix} = \begin{bmatrix} r\cos\theta & r\sin\theta \\ -r\sin\theta & r\cos\theta \end{bmatrix},$$

where $r = \sqrt{a^2 + b^2}$; thus, every similarity matrix is the product of a rotation and a scaling.

■ *A matrix is a similarity matrix if and only if (1) it has a positive determinant and (2) the linear transformation associated to it is a similarity transformation – that is, it preserves the shape of figures.*

In particular, a similarity transformation takes squares to (possibly larger or smaller) squares, which explains what we have seen in the pictures of the map $(x, y) \mapsto (x^2 - y^2, 2xy)$.

In general, a transform from 2D to 2D is called **conformal** if its Jacobian matrix is a similarity matrix at all but a few isolated points. Such a map looks like a similarity transformation in small regions and hence preserves the angles between paths even though it may distort large shapes wildly. Thus, the map $(x, y) \mapsto (x^2 - y^2, 2xy)$ is conformal except at the origin (where it doubles angles). As I mentioned before, this map was derived from the map $z \mapsto z^2$ of complex numbers. There is a huge class of similar complex-valued functions of a complex variable that are conformal.

8.3   TRANSFORMING PATHS

The technique for transforming paths involves decomposing them first and then reassembling them transformed. The following code uses `pathforall` to scan through the current path and build an array setting up the transformed path. Then it annihilates the current path with `newpath` and scans through that array to build the transformed path. The transformation is simply applied to control points to obtain new control points. Stack manipulations are used for efficiency.

```
% Argument:  the name of the transforming procedure
% It takes x y -> u v
/ctransform { load   % load the procedure onto the stack
1 dict begin
 % and now give it a local name
/f exch def
 % build an array from the current path
[
  {  % x y
    [ 3 1 roll f {moveto} ]
  }
  {  % x y
    [ 3 1 roll f {lineto} ]
  }
```

```
{    % x1 y1 x2 y2 x3 y3
     [ 7 1 roll                              % [ P1 P2 P3
         f 6 2 roll                          % [ U3 P1 P2
         f 6 2 roll                          % [ U2 U3 P1
         f 6 2 roll                          % [ U1 U2 U3
         {curveto}
     ]
  }
  {
     [ {closepath} ]
  }
  pathforall
]
  % and then replace the current path
 newpath
 {
    aload pop exec
 } forall
 end } def
```

This is generally pretty unsatisfactory unless the components of the path are small.
A line segment is just mapped onto another line, and this will usually ignore the
nonlinearity of f. For this reason it is useful to subdivide the current path one or
more times and thus to break segments into smaller ones. This is easy to do with
a routine subdivide, which replaces line segments by Bézier curves and bisects
Bézier curves into two smaller curves.

8.4 MAPS

The classical examples of 2D transformations, although with an implicit 3D twist,
occur in the design of maps of the Earth. The basic problem is that there is no
faithful way to render the surface of a sphere on a flat surface. There is no one,
single kind of map that suffices for all purposes, and various kinds must be designed
to conform to various criteria.

Coordinates on the sphere are east longitude x and latitude y. On the assumption
that the Earth's surface is a sphere of radius R, the coordinate map takes

$$(x, y) \longmapsto (R \cos x \, \cos y, \; R \sin x \, \cos y, \; R \sin y).$$

The image of a small rectangle $dx \times dy$ is an approximate rectangle in space of
dimensions $R \cos y \, dx \times R \, dy$. Of course the coordinates are singular near the
poles because they have indeterminate longitude.

I'll assume from now on for convenience that $R = 1$. (This is just a matter of choosing units of lengths correctly.) The simplest map just transforms longitude and latitude into x and y:

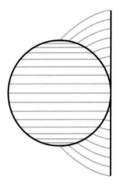

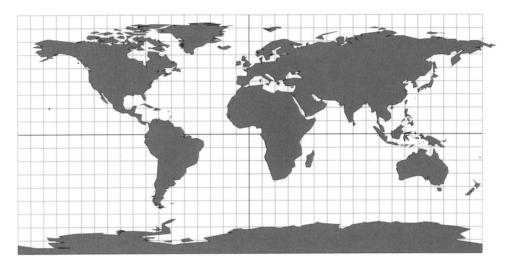

There's not much to be said for it. It preserves distance measured along meridians but wildly distorts distances along parallels.

The next simplest map is called the **cylindrical projection** because it projects a point on the Earth's surface straight out to a vertical cylinder wrapped around the Earth touching at the equator. Explicitly,

$$(x, y) \longmapsto (x, \sin y).$$

It was proven by Archimedes in his classic "On sphere and cylinder" that this map preserves areas, although it certainly distorts distances near the poles. The modern proof that area is preserved calculates that the region $\cos y\, dx \times dy$ maps to that of dimensions $dx \times \cos y\, dy$.

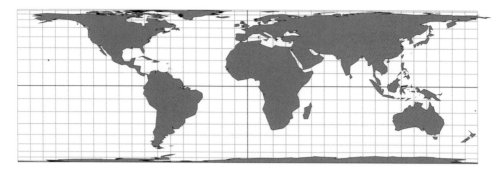

A more interesting one is the Mercator projection

$$(x, y) \mapsto \big(x, \ln\tan(\pi/4 + y/2)\big).$$

Here, distances and areas near the poles are grossly exag-
gerated, but angles are conserved, so that plotting a route by
compass is relatively simple. Indeed this is exactly the purpose
for which this projection was designed. The first map of this
sort was constructed by the sixteenth-century geographer
Gerardus Mercator himself, but it was likely the Englishman

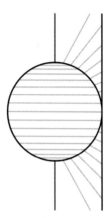

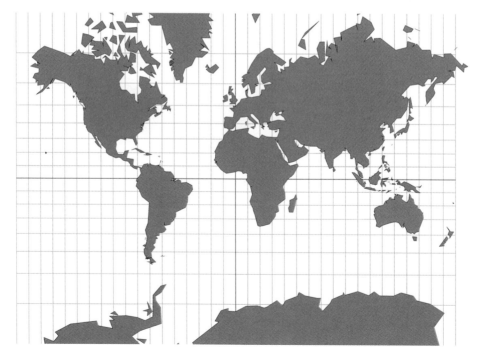

Thomas Harriot who understood its mathematical basis thoroughly – well, as thoroughly as could be done without calculus.

Why is the Mercator map conformal? Recall that the image of a small coordinate rectangle dx by dy is very approximately a rectangle on the sphere of dimensions $\cos y\, dx \times dy$. Its image under Mercator's map has size dx by

$$\tan'(\pi/4 + y/2)\, dy = \frac{dy}{\cos y}.$$

But these two rectangles are similar to each other with a scale factor of $1/\cos y$. In other words, the function $\tan(\pi/4 + y/2)$ has been chosen precisely because its derivative is $1/\cos y$.

All of the maps exhibited here were obtained from the original map data by applying `ctransform`.

8.5   STEREOGRAPHIC PROJECTION

Another map projection, **stereographic projection**, deserves a section all to itself since it plays an important role in much mathematics. In this projection, a point P on a sphere is mapped to the intersection P' of the line segment from P to the south pole with the equatorial plane.

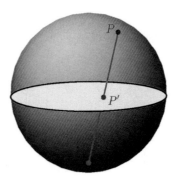

An argument using similar triangles gives us the formula

$$(x, y, z) \longmapsto \left(\frac{x}{z + 1}, \frac{y}{z + 1} \right).$$

The principal property of stereographic projection is that it is conformal. This can be proven by calculus as was the analogous result for Mercator's projection, but it can also be proven by a geometric argument going back to the English mathematician Thomas Harriot, who might have been the first to discover the result and was almost certainly the first to prove it. The argument I give and the accompanying diagrams follow Harriot's.

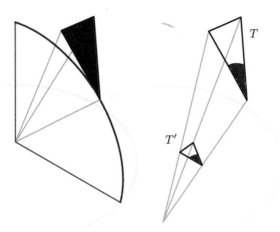

First of all, suppose P to be a given point on the sphere and P' to be its image under stereographic projection. To prove that stereographic projection is conformal, it suffices to show that any angle with respect to a meridian line on the sphere is mapped to an equal angle. Therefore, we draw at P a small right-angled triangle, one of whose sides lies tangent to the meridian from P, as in the figure at the left. The proof will show that this triangle T and its image T' under stereographic projection, shown in the figure at the upper right, are similar.

To see why this is so, construct a third triangle T'' sharing a common side with the one whose vertex is P but parallel to the projected triangle with a vertex P'' on the line through the south pole, P', and P. It is clearly similar to the projected triangle T', and thus to show that T and T' are similar it suffices to show that T'' is congruent to T.

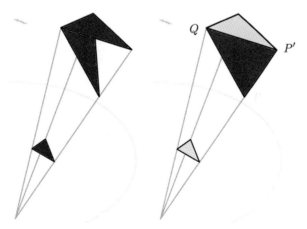

The two triangles T and T'' share a common side and are both right-angled at Q. Therefore, a side–angle–side argument will show that they are congruent once it is known that the sides PQ and $P'Q$ have the same length. This is because the meridian triangle PQP'' is isosceles, and this can be seen from an elementary two-dimensional diagram:

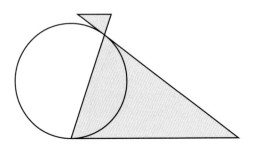

8.6 **FONTS WANT TO BE FREE**

Text can be transformed, too. After setting up a font and defining procedures `subdivide` and `ctransform`,

```
newpath
1 0 moveto
(Roman) true charpath
subdivide
/f ctransform
gsave
1 0.7 0.7 setrgbcolor
fill
grestore
currentlinewidth 2 div setlinewidth
stroke
```

produces (except for the comparison image in light gray)

Subdivision is important here because font paths possess many straight segments that are much better off transformed as curves.

8.7 **CODE**

The procedures `subdivide` and `ctransform` are to be found in the package `trans-form.inc`. The map data can be found in the PostScript files `coasts.1.inc` etc,

which I have made from the MWDB files now in the World Database. The index
indicates level of detail; index 1 is the greatest detail, index 5 the least.

REFERENCES

1. J. L. Berggren and A. Jones, **Ptolemy's Geography**, Princeton University Press,
    1997. Mathematical map making began with the Greeks. The illustrations
    here, many taken from medieval manuscripts, are beautiful. The invention of
    printed books in the West meant not only a wider audience for science but
    also, ironically, a loss in the quality of graphics and in particular the loss of
    brilliant color.

2. Timothy G. Freeman, **Portraits of the Earth – A Mathematician Looks at Maps**,
    American Mathematical Society, 2002. An appendix includes an account of
    how what seems to be the same map data that I have used can be used to draw
    maps using Maple™.

3. David Hilbert and Stephan Cohn-Vossen, **Geometry and the Imagination**,
    Chelsea, 1952. Many maps are conformal – preserve angles – and one of
    the basic theorems in the subject is that stereographic projection is a con-
    formal map. Nowadays this can be proven quickly by calculus, but an ele-
    gant geometric proof can be found in Section 36 of this book. Early Greek
    astronomers knew that stereographic projection maps circles to circles, ap-
    plying a theorem of Apollonius on conic sections to deduce it. This fact is
    closely related to conformality, but conformality itself was apparently first
    stated and proved by the English mathematician Thomas Harriot, who was
    among other things Sir Walter Raleigh's navigation expert, in about 1590.
    This was still in the golden age of map making stimulated by the discovery
    and exploration of America as well as the great Portugese voyages around
    Africa to the Far East. Harriot's proof was unpublished, but it can be re-
    constructed from a rather handsome sketch in his manuscripts. A repro-
    duction can be found in the article by J. A. Lohne, "Thomas Harriot als
    Mathematiker," Centaurus 11 (1965/66), who also discusses Harriot's treat-
    ment of the Mercator projection. The first published proof was by Edmund
    Halley, who is often (and incorrectly) given credit for it. His paper is "An Easie
    Demonstration of the Analogy of the Logarithmic Tangents . . . " in the Philo-
    sophical Transactions 19 (1695–1697), pp. 202–214 (miraculously available
    through the electronic library service JSTOR). Although he acknowledges that
    he learned the statement from De Moivre, Halley claims the proof as his own,
    but there is suspicion that both came one way or another from the remark-
    able Harriot (as did probably much other mathematics of the seventeenth
    century).

4. John McCleary, **Geometry from a Differentiable Viewpoint**, Cambridge Uni-
    versity Press, 1994. There is an interesting account of map projections in

Chapter 8 which, like much of this book, would serve nicely as a source of graphics projects.

5. Tristan Needham, **Visual Complex Analysis**, Oxford University Press, 1997. This explains the relationship between complex numbers and conformal maps with many illustrations.

6. John P. Snyder, **Flattening the Earth: Two Thousand Years of Map Projections**, University of Chicago Press, 1993. This is a very readable history of map making, including descriptions of dozens of different ones.

7. For the maps I have used the data from the "World Database II," which is now

    included in the file `world.zip` accessible by clicking on the icon

    at

    `http://archive.msmonline.com/1999/12/vis2.htm`

    These map data, which are now in the public domain, were compiled by Fred Pospeschil and Antonio Riveria from data originally created by the Central Intelligence Agency. The particular files I have used incorporate further modifications by Paul Anderson of Global Associates, Ltd. The Bodleian Library of Oxford University maintains a convenient Web page with links to sources of other map data available without cost.

CHAPTER 9

Recursion in PostScript

For various technical reasons, recursion in PostScript is a tricky business. It is possible, however, and in some circumstances nearly indispensable. It is extremely useful to have some idea of how to deal with recursion because standard algorithms in the literature, for example those traversing lists of various kinds, are often laid out in terms of recursion.

9.1 THE PERILS OF RECURSION

The factorial $n!$ of a positive integer n is defined informally by the rule that it is the product of all positive integers less than or equal to n. Thus $4! = 1 \cdot 2 \cdot 3 \cdot 4 = 24$. If we want to get a computer to calculate it, we follow these rules: (1) if $n = 1$ then $n! = 1$; (2) if $n > 1$ then $n! = n \cdot (n-1)!$. This sort of formula, where a function is evaluated directly by reduction to a simpler case, is called **recursion**. It could be argued that recursion encapsulates the essence of mathematics, which tries to reduce every assertion either to an axiom or to one that has already been proven or in broad terms to reduce every problem to a simpler one that has already been solved.

At any rate, we can write a procedure in PostScript that follows these two last rules exactly:

```
/factorial { 1 dict begin
  /n exch def
  n 1 eq {
    1
  }{
    n n 1 sub factorial mul
  } ifelse
end } def
```

This code is correct and will run but probably only if n is small. Why is that? The reason for failure is somewhat technical. When you begin a dictionary in a procedure, that dictionary is put on top of a stack of dictionaries, and when you end that dictionary it is taken off this stack. Unlike a few other stacks used by PostScript, the dictionary stack is usually severely limited in size. If you call a procedure recursively that uses a local dictionary, the size of the dictionary stack will build up with every call – quite possibly beyond the maximum size allowed. If this occurs you will get a dictstackoverflow error message. So this procedure might very well work for small values of n but fail for large ones.

Therefore

■ *You should never introduce a dictionary to handle local variables in a recursive procedure exactly as you do in others.*

There is nothing wrong with using dictionaries in recursive procedures, but *they shouldn't remain open across the recursive procedure calls.* That is to say, you should begin and end the dictionary without making any recursive calls in between. You might very well want to do this so as to do some complicated calculations before setting up the recursion calls. We'll see an example later on. The simplest rule to follow with recursive procedures in PostScript is to use variables as little as possible and to resort to stack manipulations instead. In effect, you should use data on the stack to serve as a substitute for the execution stack PostScript doesn't want you to use. The drawback, of course, is that such manipulations are very likely to make the procedure much less readable than it should ideally be. The technique is usually painful and bug-prone and hence is best avoided. Nonetheless, here is a simple example – a correct recursive procedure to calculate $n!$ correctly (except for problems with floating point accuracy):

```
% args: a single positive integer n
% effect: returns n!
/factorial {
  % stack: n
  1 gt {
    % stack: n > 1
    % recall that n = n(n-1)!
    dup 1 sub
    % stack: n n-1
    factorial
    % stack: n (n-1)!
    mul
```

```
      } ifelse
       % stack: n!
    } def
```

The comments here trace what is on the stack, which is an especially good idea in programs in which complicated stack manipulations are made.

Of course this version of `factorial` is hardly efficient. It is simple enough to write a `factorial` routine that just uses a `for` loop. But this version illustrates nicely the perils of recursion in PostScript.

EXERCISE 9.1. *The gcd (greatest common divisor) of two nonnegative integers m and n is n if m = 0 or the gcd of m and m* mod *m if m ≠ 0. Construct a recursive procedure in PostScript to find it.*

Overflow on the dictionary stack occurs in my current implementation only when the stack reaches a size of about 500. So much of what I am saying here is silly in many situations. But there is another reason not to allow the dictionary stack to grow large: when looking up the value of a variable, the PostScript interpreter looks through all of the dictionaries on the stack, starting at the top, until it finds the variable's name as a key. If the name is inserted in one of the bottom dictionaries, much searching has to be done. The efficiency of a recursive procedure can thus be cut down dramatically as the recursion builds up.

9.2   SORTING

A common and extremely useful example of a procedure that uses recursion is a sorting routine called **quick sort**. I am going to present it below in this section without much explanation, but I will make a few preliminary remarks.

Let me first explain what just about any sorting routine will do. It will have one argument, an array of items that can be ranked relative to each other – for example an array of numbers. The routine will rearrange the items in the array in ascending order, from smallest up. For example

```
    [4 3 2 1] quicksort
```

will rearrange the items in the array to make it equal to [1 2 3 4]. You might think that this isn't a very mathematical activity and that a mathematician would have no serious interest in such a routine – especially in PostScript, where you are only interested in drawing stuff. That is not at all correct. *A good sorting routine should be part of the tool kit of every mathematical illustrator.* For example, we shall see a bit later a procedure that constructs the convex hull of a given finite collection of points and that depends on a sorting routine in a highly nontrivial way. Sorting routines can also play a role in 3D drawing, where objects must be drawn from back to front.

As any introduction to programming will tell you, there are several ways to sort items in an array. They vary enormously in efficiency. One very simple one is called the **bubble sort** because it pushes big items up to the top end of the array as though they were bubbles. It makes several scans of the array, each time bubbling up the largest item. For example, on the first pass bubble sort moves the largest item to the end, on the second it moves the second largest item to the place next to the end, and so on.

Here is PostScript code for a bubble sort.

```
% args:  array a of numbers
% effect:  sorts the array in order
 /bubblesort { 4 dict begin
 /a exch def
 /n a length 1 sub def
 n 0 gt {
    % at this point only the n+1 items in the bottom of a remain to be sorted
    % the largest item in that block is to be moved up into position n
    n {
      0 1 n 1 sub {
        /i exch def
        a i get a i 1 add get gt {
          % if a[i] > a[i+1] swap a[i] and a[i+1]
          a i 1 add
          a i get
          a i a i 1 add get
          % set new a[i] = old a[i+1]
          put
          % set new a[i+1] = old a[i]
        put
      } if
    } for
    /n n 1 sub def
  } repeat
} if
end } def
```

For example, if we sort [5 4 3 2 1 0] in this way, we get successively in each bubble

```
[4 3 2 1 0 5]
[3 2 1 0 4 5]
[2 1 0 3 4 5]
[1 0 2 3 4 5]
[0 1 2 3 4 5]
[0 1 2 3 4 5]
```

Bubble sorting is very easy to put into correct PostScript code. But it is pretty inefficient if the size n of the array is large. On the first bubbling pass it makes $n - 1$ comparisons, on the second $n - 2$, and so on. This makes approximately $n(n - 1)/2$ comparisons in all, and so the time it takes is proportional to the square of n.

We can do much better. The sorting routine, which is generally fastest of all, is called **quick sort**. One of the things that makes it unusual is that it tries a **divide and conquer** approach basically splitting the array into halves and then calling itself recursively on each half. Its running time is nearly always proportional to $n \log n$, which is a great improvement over bubblesort.

Quick sort has three components. The principal routine is quicksort itself with only the array as an argument. It calls a routine subsort that has as argument not only the array but also two indices $L < R$, picking out a range inside the array to sort. For the call in quicksort these are 0 and $n - 1$. But subsort calls itself recursively with smaller ranges. It also calls a third routine partition, which does the real swapping. This procedure has four arguments – the array a, a lower index L and an upper index R, and an integer x. It moves items in the $[L, R]$ range of the array around so the left half $[L, i]$ is made up of items less than or equal to x and the right half $[j, R]$ of items greater than or equal to x. It then returns the pair i and j on the stack. I hide the details of partition, but here is the pseudocode for subsort:

```
subsort(a, L, R) {
  x := a[(L+R)/2];
  [i, j] = partition(a, L, R, x);
  if (L < j) subsort(a, L, j);
  if (i < R) subsort(a, i, R);
}
```

and here in PostScript is the whole package:

```
% args:  a L R x
% effect:  effects a partition into two pieces [L j] [i R]
% leaves i j on stack
/partition { 1 dict begin
... end } def

% args:  a L R
% effect:  sorts a[L .. R]
/subsort {
  1 dict begin
  /R exch def
  /L exch def
  /a exch def
  % x = a[(L+R)/2]
  /x a L R add 2 idiv get def
  a L R x partition
  /j exch def
  /i exch def
  % put recursion arguments on the stack
  % as well as the arguments for the tests
  a L j
  j L gt
  a i R
  i R lt
  % close dictionary
  end
  { subsort }{
    % get rid of unused arguments
    pop pop pop
  } ifelse
  { subsort }{ pop pop pop } ifelse
} def

% args:  a
% effect:  sorts the numerical array a
% and leaves a copy on the stack
/quicksort { 4 dict begin
  /a exch def
  /n a length 1 sub def
```

```
      n 0 gt {
        a 0 n subsort
      } if
      a
    } def
```

The important thing here is to notice how the recursive routine `subsort` manages the dictionary and stack in a coordinated way. The most interesting thing in the way it does that is how the arguments for the recursion are put on the stack before it is even known if the calls will be made, the arguments are removed if not used. The procedure `partition`, on the other hand, does not call itself and is therefore allowed to use a local dictionary in the usual way.

This will give you an idea of what goes on, but the working version of the `quicksort` routine has an extra feature. We will be interested in sorting arrays of things other than numbers in a moment. Rather than design a different sort routine for each kind of array, we add to this one an argument /comp, the name of a comparison test replacing `lt` for numbers. For numbers themselves, we would define

```
    /comp { lt } def
```

and then write

```
    /comp [5 4 3 2 1 0] quicksort
```

This is an extremely valuable modification.

To get an idea of how much faster `quicksort` is than `bubblesort`, consider that if $n = 1,000$, then $n(n-1)/2 = 495,000$, whereas $n \log n$ is about $7,000$. Of course each loop in `quicksort` is more complicated. In actual runs, `quicksort` takes 0.07 seconds to sort $1,000$ items, whereas `bubblesort` takes 0.79 seconds on the same array. The `quicksort` is definitely fast enough to include in drawing programs without noticeable penalty.

EXERCISE 9.2. *Use recursion to draw the following picture with an optional parameter to determine the depth of the recursion:*

Notice that even the line widths get smaller farther out along the branches. If you are a mathematician, you might like to know that this is the Bruhat–Tits building of $SL_2(\mathbb{Q}_2)$. If you are not a mathematician, you can think of it as one nonmathematician of my acquaintance called it: "chicken wire on growth hormones."

9.3   CONVEX HULLS

Mostly to convince you that sorting is not an empty exercise but also to give you a procedure you might find very useful, I describe here a well-known and relatively efficient procedure that finds the convex hull of a finite set of points in the plane. It will have a single argument, an array of points, which we may as well take to be arrays [x y]. It will return an array made up of a subset of the given set of points listed so that traversing them in order will go clockwise around the convex hull of the point set. The output, in other words, consists of the **extreme** members of the original set, those which are in some sense on its outside. Finding the outside points, and in the correct order, is not exactly a difficult task, but before you start the discussion you might think of how you would do it yourself. Extra complications are caused by possible errors in the way PostScript handles floating point numbers.

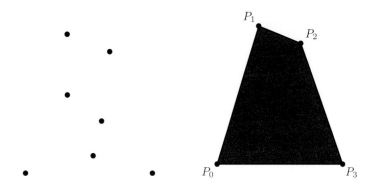

The algorithm to be described is not in fact terribly complicated, but certain aspects of it might not seem so well motivated without a few preliminary remarks. First of all, what does sorting have to do with the problem? The picture at the right illustrates that points of the original data set farthest to the left and farthest to the right will be part of the output.

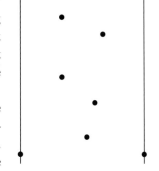

But finding these points will involve sorting the original point set according to values of the x coordinate. Indeed, we have to be somewhat more careful. It might very well be the case that several points in the set have the same x coordinate, and so there might not be a unique point farthest left or right. We will therefore be more particular: we say that (x_0, y_0) is less than the point (x_1, y_1) if $x_0 < x_1$ (the first point is farther left) or $x_0 = x_1$ but $y_0 < y_1$ (in case they have equal x coordinates, the one on the bottom will be called smaller). When we sort the data, this is the ranking procedure we use. It is still true that the largest and smallest points in the set must be part of the output.

In fact, sorting the original points is one of two key steps in the procedure. We may in effect assume that the original set of points is an array in which the points are listed so that those that are "smaller" in the sense described in the preceding paragraph come first. This immediately reduces the difficulty of the problem somewhat since we can take the first point of the (sorted) input to be the first point in the output.

In the figure at the right, the points are linearly ordered along a polygon "from left to right" in the slightly generalized sense.

Another idea is that we will do a left-to-right scan of the input, producing eventually all the points in the output that lie along the top of the convex hull. Then we will do a right-to-left scan that produces the bottom. Again, the preliminary sorting guarantees that the last point in the array will end the top construction and begin the bottom one.

What is it exactly that we are going to do as we move right? At any moment we will be looking at a new point in the input array. It will either have the largest x-value among all the points looked at so far or it will at least have the largest y-value among those that have the largest x-value. In either event, let its x-value be x_r. Then we will try to construct the top half of the convex hull of the set of all input points with x-value at most x_r. Certainly the current point P we are looking at will be among these. But as we add it, we might very well render invalid a few of the points on the list we have already constructed. We therefore go backwards among those, deleting from the list the ones that lie distinctly inside the top hull of the whole list. A point P_k will be one of those precisely when the polygon $P_{k-1} P_k P$ has an upward bend and thus the polygon $P_{k-1} P_k P$ is below the segment $P_{k-1} P$. So we go backwards, throwing away those in such a bend until we reach a triple where the bend is to the right. Then we move on to the next point to the right. The following figures show the progression across the top in a particular example, including the "left turns" that are eliminated as we go along:

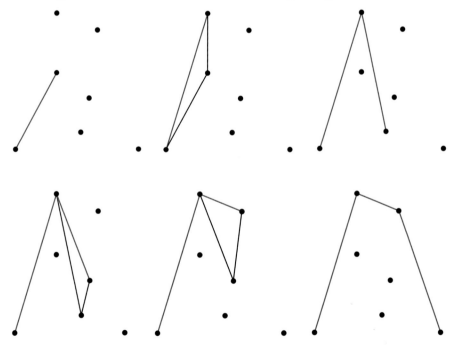

I reproduce below the principal routine for finding the convex hull. It uses routines comp for sorting, vsub, which subtracts two vectors, dot to calculate the dot product of two vectors, and vperp, which rotates the difference of two vectors counterclockwise by 90°. The routine /comp tests whether $(x_0, y_0) < (x_1, y_1)$. All these are defined internally in hull.inc.

```
% args:   an array of points C
 % effect:  returns the array of points on the boundary of
 % the convex hull of C, in clockwise order
/hull { 16 dict begin
  /C exch def
   % sort the array
   /comp C quicksort
   /n C length def
     % Q circles around to the start
     % make it a big empty array
   /Q n 1 add array def
   Q 0 C 0 get put
   Q 1 C 1 get put
   /i 2 def
   /k 2 def
     % i is next point in C to be looked at
     % k is next point in Q to be added
     % [ Q[0] Q[1] ... ]
     % scan the points to make the top hull
   n 2 sub {
       % P is the current point at right
      /P C i get def
      /i i 1 add def
      {
          % if k = 1 then just add P
         k 2 lt { exit } if
          % now k is 2 or more
          % look at Q[k-2] Q[k-1] P: a left turn (or in a line)?
          % yes if (P - Q[k-1])*(Q[k-1] - Q[k-2])perp >= 0
         P Q k 1 sub get vsub
         Q k 1 sub get Q k 2 sub get vperp
         dot 0 lt {
             % not a left turn
           exit
         } if
```

```
            % it is a left turn; we must replace Q[k-1]
            /k k 1 sub def
        } loop
        Q k P put
        /k k 1 add def
    } repeat
     % done with top half
     % K is where the right hand point is
    /K k 1 sub def
    /i n 2 sub def
    Q k C i get put
    /i i 1 sub def
    /k k 1 add def
    n 2 sub {
        % P is the current point at right
        /P C i get def
        /i i 1 sub def
        {
            % in this pass k is always 2 or more
            k K 2 add lt { exit } if
            % look at Q[k-2] Q[k-1] P: a left turn (or in a line)?
            % yes if (P - Q[k-1])*(Q[k-1] - Q[k-2])perp >= 0
            P Q k 1 sub get vsub
            Q k 1 sub get Q k 2 sub get vperp
            dot 0 lt {
                % not a left turn
                exit
            } if
            /k k 1 sub def
        } loop
        Q k P put
        /k k 1 add def
    } repeat

    % strip Q down to [ Q[0] Q[1] ... Q[k-2] ]
    % excluding the doubled initial point
    [ 0 1 k 2 sub {
      Q exch get
    } for ]
end } def
```

9.4 CODE

The code for sorting is in `sort.inc`; that for finding convex hulls is in `hull.inc`.

REFERENCES

1. M. de Berg, M. van Creveld, M. Overmars, and O. Schwarzkopf, **Computational Geometry – Algorithms and Applications**, Springer-Verlag, 1997. The convex hull algorithm is taken from Chapter 1 of this book. Like many books on the subject of computational geometry, the routines it explains are often more beautiful than practical. Among them, however, are a few useful gems. And the illustrations are absolutely first rate, which is rarely the case in computer graphics books.

2. R. Sedgewick, **Algorithms**, Addison-Wesley. There are several editions of this text, complementary algorithms in various programming languages. The latest one (2003) uses Java. Chapter 5 has a nice discussion of recursion and tree structures, and Chapters 6 and 7 are about sorting.

## CHAPTER 10

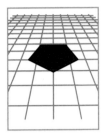

Perspective and homogeneous coordinates

Many image manipulation programs have a tool for modifying the perspective of images. Thus, I start out with this familiar picture and then I import it into my image manipulation program (which happens to be the **G**nu **I**mage **M**anipulation **P**rogram or **GIMP**). I next open the transform tool and choose the perspective option. When I click now in the image window, what I see is this:

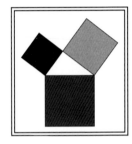

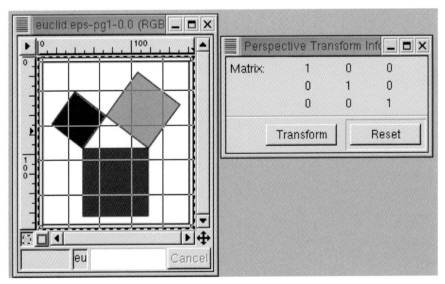

The tool lays down a grid on top of my image and simultaneously brings up what it calls a *Perspective Transform Info* window in which is displayed a 3×3 matrix. It also shows some boxes at the corners of the grid that I can grab and move around. When I do so, the original image remains the same, and the grid adjusts itself to my choices. In effect, a geometrical transform of some kind is applied to the grid. The matrix changes, alas, in some apparently incomprehensible manner.

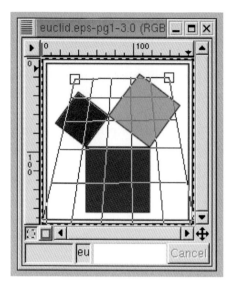

Finally, if I click on the **transform** button in the matrix window, the image itself is transformed:

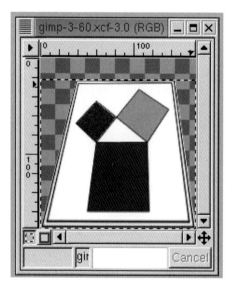

The original image has been distorted in such a way as to make it look as though it were being viewed from below – as though it were being viewed in a different **perspective**.

What kind of a transform are we seeing here? How is it different from others we have already seen such as translation, rotation, shearing, and scale change? The image manipulation program has in fact other transform tools corresponding to those such as the **rotate** and **scale** tools. All those other transforms have one property that the perspective transform doesn't: they all transform a pair of parallel lines into some other pair of parallel lines. They are all affine transforms, and all affine transforms have this property. So the perspective transform is really something new. As we'll see, in gaining flexibility we must sacrifice something else.

This chapter will analyze perspective transforms mathematically. The matrix window in the figures above turns out to be a major clue. The analysis will involve at first what seems like an enormous detour. Before we begin on that journey, I should motivate it by mentioning that perspective in 3D is a crucial ingredient in any graphics toolkit and that the techniques to be explained here are, for that reason, ubiquitous. The advantage of working with this 2D image manipulation tool is that things are simpler to visualize. It should be thought of as a model for the more interesting 3D version.

10.1 THE PROJECTIVE PLANE

To understand what the perspective tool does we must enlarge the usual plane by adding points "at infinity." Familiarity with the use of perspective in art, where lines meet at infinity and the horizon becomes a line on a canvas, should make this seem like a natural step. We have seen already a hint that this is necessary: two parallel lines meet only at infinity, but after a perspective transform is applied to them they may meet in a definite point of the image plane.

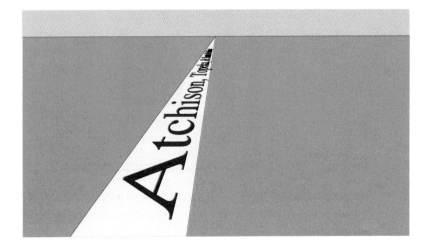

This means that the transform has converted some points at infinity into ordinary ones. We will see also that it may move some ordinary points off to infinity as well.

How can the plane be enlarged by adding points at infinity? There are several ways to do this. The one I'll follow is not quite the obvious one. I'll add one point at infinity for every **nonoriented direction**. In other words, on every line there exists one of these new infinite points, and we can reach it, so to speak, going infinitely far along the line – *in either direction*. The more obvious thing to do would be to add a point at infinity for every direction so that every line contains two points at infinity, but this turns out not to be so useful.

The extended plane we get by adding points at infinity to the usual plane in this way is called the **projective plane**. We can put coordinates on its points. There are two cases to be dealt with: (1) The ordinary finite points can be specified by the usual coordinates (x, y). (2) A direction can be specified by a vector $[x, y]$, but x and y are only determined up to a nonzero scalar multiple; that is, $[cx, cy]$ has the same direction as $[x, y]$. Because of this ambiguity, the second pair of coordinates are called **homogeneous coordinates**.

Any two parallel lines have the same direction, and so they will contain the same unique point at infinity. In other words, with these new points added every pair of parallel lines will intersect at a single point. In fact, every two distinct lines, whether parallel or not, intersect in exactly one point. This is a very simple property of the new plane.

Adding the points at infinity in the way I have done it is intuitive if somewhat awkward. In particular it doesn't seem to give any clues about what GIMP does. And it is bothersome that the points at infinity are treated differently from the rest. There is a surprisingly elegant but equivalent way to carry out the construction.

We have already seen how to embed 2D points (x, y) into 3D by adding a third coordinate $z = 1$ to get $(x, y, 1)$. The ordinary 2D plane $z = 0$ becomes the plane $z = 1$. But it turns out that we can also embed the points at infinity into 3D? A direction $[x, y]$ in 2D is to be identified with the same direction $[x, y, 0]$ in the plane $z = 0$.

Thus far, the finite points and the infinite ones are still treated quite differently. But in fact *we can identify even the points in the plane $z = 1$ with directions in 3D*. A direction in 3D is essentially a vector $[x, y, z]$ (with one coordinate not 0), and this determines a unique line through the origin with that direction: that of all points (tx, ty, tz) as t ranges from $-\infty$ to ∞. That line will intersect the plane $z = 1$ in a unique point $(x/z, y/z, 1)$ (with $t = 1/z$) as long as $z \neq 0$. If z vanishes, then the line with direction $[x, y, 0]$ amounts to a point at infinity on the projective plane. In summary,

■ *The points of the extended plane we get by adding unoriented directions to the usual plane are in exact correspondence with lines in 3D through the origin.*

To repeat: the 3D line with direction $[x, y, 0]$ corresponds to the direction $[x, y]$ in 2D, a point at infinity, and the line with direction $[x, y, z]$, where $z \neq 0$, corresponds to the "finite" point $(x/z, y/z)$.

A line through the origin in 3D may be assigned the vector $[x, y, z]$ indicating its direction. These are **homogeneous coordinates** in the sense that multiplying them all by a nonzero constant doesn't change the direction. If $c \neq 0$, then the coordinates $[x, y, z]$ and $[cx, cy, cz]$ are associated to the same point of the projective plane. We have seen homogeneous coordinates before when we assigned the coordinates $[A, B, C]$ to the line $Ax + By + C = 0$.

If (x, y, z) is any point in 3D with $z \neq 0$, then the point $(x/z, y/z, 1)$ on our copy of the usual 2D plane that is equivalent to it will be called its **projective normalization**. This corresponds in turn to the 2D point $(x/z, y/z)$, which I call its **2D projection**.

There is one important relationship between the ordinary points and points at infinity. If we start at a point $P = (a, b)$ and head off to infinity in the direction $[x, y]$, then we pass through points $P + t[x, y] = (a + tx, b + ty)$. As t gets larger and larger, we should be approaching some point at infinity, shouldn't we? Sure enough, in terms of homogeneous coordinates we are passing through the points

$$[a + tx, b + ty, 1] \sim [a/t + x, b/t + y, 1/t] \longrightarrow [x, y, 0]$$

as $t \to \infty$.

10.2   BOY'S SURFACE

The projective plane has one very surprising property. If we move along a line to infinity we will pass through infinity and come back towards where we started – but from the opposite direction. We can see this in a simple **topological** model of the projective plane – that is, one that preserves its qualitative features. In this model a point on the boundary of the square is identified with its opposite. If we follow this procedure with an oriented square (i.e., pass it off through infinity and back again from the other direction), it comes back oriented the other way! This is shown in the following figures in the topological model:

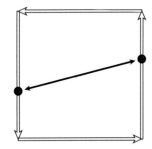

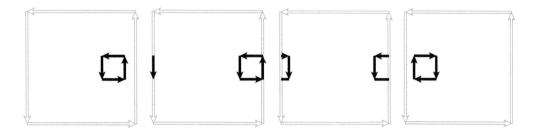

After a little thought, this makes sense. The opposite sides are identified but with a twist much like that of the familiar Möbius strip. The projective plane is **nonoriented**, one-sided just like the Möbius strip.

There is still something unsatisfactory about my definition of the projective plane. If it really is a 2D surface, why can't we picture it as we do a sphere or a doughnut? Because it is nonorientable. Like the Klein bottle, which is also one-sided, it can only be realized in 3D as a surface that intersects itself. It is called **Boy's surface** after its discoverer, Werner Boy. Here are the original sketches (from his 1903 *Mathematische Annalen* article) showing how to construct a qualitatively correct version of it.

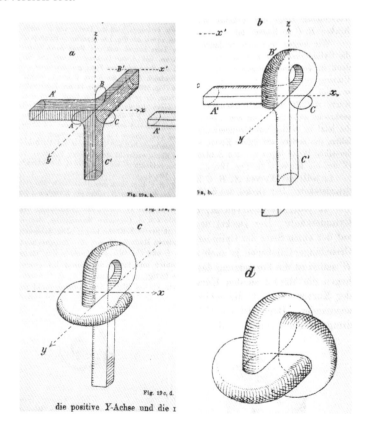

It is an interesting exercise to walk around on the surface and verify that it is one-sided. To see why it is a topological model for the projective plane is more difficult, and I won't really attempt it. To begin to understand what's at stake, keep in mind that each line through the origin in 3D intersects the unit sphere in exactly two opposite or **antipodal** points such as (x, y, z) and $(-x, -y, -z)$. Thus, in modeling the projective plane as a surface, we are looking for a way to map the sphere onto a surface in which these two antipodal points collapse to one.

10.3   PROJECTIVE TRANSFORMATIONS

The payoff in this business is that we can enlarge the collection of transformations that can be applied to the points of the plane. We have seen that PostScript implicitly conceives of the affine transformation

$$x' = ax + cy + e$$
$$y' = bx + dy + f$$

in terms of the 3D linear transformation

$$[\, x' \;\; y' \;\; 1\,] = [\, x \;\; y \;\; 1\,] \begin{bmatrix} a & b & 0 \\ c & d & 0 \\ e & f & 1 \end{bmatrix}.$$

This works because the 3D linear transformation corresponding to this particular matrix takes the plane $z = 1$ into itself. We now enlarge the transformations of the plane $z = 1$ by allowing right multiplication by an arbitrary 3×3 matrix that happens to be invertible (i.e., with nonzero determinant). Such a linear transformation takes lines through the origin into other lines through the origin and therefore acts as a transformation of points of the projective plane – a **projective transformation**. It will not often take points with $z = 1$ into other points with $z = 1$, nor even necessarily into other points with $z \neq 0$, but it will usually do so. What does this mean in algebraic and geometrical terms?

Any 3×3 matrix acts as a transformation on the usual Euclidean space of three dimensions. Suppose T to be a 3×3 matrix and (x, y) a 2D point. How can T be used to transform this 2D point into another 2D point?

(1) Extend it to a 3D point by adding on 1 as a last coordinate, and so it becomes

   $(x, y, 1)$.

(2) The matrix T will transform this to another 3D point

   $(x, y, 1)\, T = (x', y', z')$.

(3) To get a 2D point back again, divide through by z' and throw away the final coordinate to get

$$(x'/z', y'/z').$$

There will be problems when $z' = 0$, and we will discuss these Section 10.5.

10.4 THE FUNDAMENTAL THEOREM

There is one fundamental theorem in this subject that makes the GIMP perspective tool practical. We have seen that the tool allows the user to move any one of four points around with the mouse to obtain the transformation desired. The mathematical assertion to go with this is

■ *Consider four points P, Q, R, S in the projective plane with the property that no three of them lie on any one line. If P', Q', R', S' are another four points with the same property, then there exists a unique projective transformation taking P to P' and so forth.*

I'll come back to this in a while but place it in context first. Here is an analogous but simpler result:

■ *Given three vectors u, v, w in 3D that are linearly independent. If u', v', w' are any other three vectors, then there exists a unique 3×3 matrix transformation taking u to u' and so forth.*

In the special case in which u, v, and w are the three basis vectors

$$e_1 = [1, 0, 0]$$
$$e_2 = [0, 1, 0]$$
$$e_3 = [0, 0, 1]$$

the matrix can be written down explicitly. It is

$$\begin{bmatrix} u' \\ v' \\ w' \end{bmatrix},$$

whose rows are u', v', w'. In the general case, let T take e_1 to u and T' take e_1 to u' and so on.

$$\begin{bmatrix} u \\ v \\ w \end{bmatrix} \xleftarrow{\ T\ } \begin{bmatrix} e_1 \\ e_2 \\ e_3 \end{bmatrix} \xrightarrow{\ T'\ } \begin{bmatrix} u' \\ v' \\ w' \end{bmatrix}$$

Then the transformation $T^{-1}T'$ with matrix

$$\begin{bmatrix} u \\ v \\ w \end{bmatrix}^{-1} \begin{bmatrix} u' \\ v' \\ w' \end{bmatrix}$$

takes u to u' and so on.

Here is a consequence:

■ *Given three points P, Q, R in the 2D plane, no two of which lie in a single line. If P', Q', R' are any another three 2D points, then there exists a unique affine transformation taking P to P' and so on.*

Embed these 2D points in 3D by making $z = 1$. According to the previous result, the 3D linear transformation whose matrix is

$$\begin{bmatrix} x_P & y_P & 1 \\ x_Q & y_Q & 1 \\ x_R & y_R & 1 \end{bmatrix}^{-1} \begin{bmatrix} x_{P'} & y_{P'} & 1 \\ x_{Q'} & y_{Q'} & 1 \\ x_{R'} & y_{R'} & 1 \end{bmatrix}$$

takes P to P' and so forth. It is easy enough to check that it must then take the whole plane $z = 1$ into itself. From this it follows But then it can be checked in turn that this matrix has the form

$$\begin{bmatrix} * & * & 0 \\ * & * & 0 \\ * & * & 1 \end{bmatrix},$$

which means that it is the matrix of an affine transformation.

EXERCISE 10.1. *Prove the last two claims: (1) If a linear transformation takes three nonlinear points on the plane $z = 1$ into other points on the same plane, then it takes all points on the plane $z = 1$ into points on the plane $z = 1$; (2) if a linear transformation takes the plane $z = 1$ to itself, then it derives from an affine 2D transformation.*

Now let's return to the original claim about projective transformations. If we translate this claim through the definition of projective points, the following assertion results:

■ *Consider four vectors P, Q, R, S in 3D such that no three of them lie in one plane. If P', Q', R', S' are another four vectors with the same property, then there exists an invertible 3×3 matrix taking P to a scalar multiple of P', and this is true for the other vectors as well. It is unique up to a single scalar multiplication.*

In effect, the transformation takes the line through P to that through P', and so on. The proof will construct the matrix explicitly. The basic idea is similar to that of the first result in this section. I first look at the special case

$$P = e_1 = [1, 0, 0]$$
$$Q = e_2 = [0, 1, 0]$$
$$R = e_3 = [0, 0, 1$$
$$S = e_{123} = [1, 1, 1].$$

This special case is simple. Let M be the matrix with rows equal to P', Q', R':

$$M = \begin{bmatrix} P' \\ Q' \\ R' \end{bmatrix}.$$

By assumption on the vectors P' and the others the matrix will be invertible. And it takes $[1, 0, 0]$ to P', $[0, 1, 0]$ to Q', and $[0, 0, 1]$ to R', as you can calculate. By assumption also we can write

$$S' = c_{P'} P' + c_{Q'} Q' + c_{R'} R'.$$

We can find these coefficients by solving the 3×3 system of linear equations with coefficient matrix

$$\begin{bmatrix} P' \\ Q' \\ R' \end{bmatrix}.$$

The matrix

$$\begin{bmatrix} c_{P'} P' \\ c_{Q'} Q' \\ c_{R'} R' \end{bmatrix}$$

will have the same effect on the projective points $(1, 0, 0)$ and the others as M and will take $[1, 1, 1]$ to S'.

In the general case, we can find transformations T taking e_1 to a multiple of P and T' taking e_1 to a multiple of P', and the same holds true for the other vectors. But then the composite $T^{-1}T'$ takes P to a multiple of P' and so on.

10.5 PROJECTIVE LINES

A projective 2D point is now understood to be a row vector $[x, y, z]$, where x as well as y and z are determined only up to a common nonzero-scalar multiple and are not all equal to 0. If $[A, B, C]$ is another similar triple, then the set of points where

$$Ax + By + Cz = 0$$

is a plane – in fact a union of lines through the origin in 3D – and is called a **projective line**. Its intersection with the plane $z = 1$ is the familiar 2D line $Ax + By + C = 0$. We will consider projective lines as column vectors, and so $Ax + By + Cz$ is a matrix product.

Normally, the equation $Ax + By + C = 0$ with $A = B = 0$ makes no sense. But now in the projective plane, it becomes the plane $Cz = 0$ or just $z = 0$. It is the set of projective points at infinity, which is often called the **line at infinity** or in effect the horizon.

If

$$T = \begin{bmatrix} t_{1,1} & t_{1,2} & t_{1,3} \\ t_{2,1} & t_{2,2} & t_{2,3} \\ t_{3,1} & t_{3,2} & t_{3,3} \end{bmatrix}$$

is a projective transformation, it acts on 2D points by taking

$$(x, y) \mapsto \left(\frac{t_{1,1}x + t_{2,1}y + t_{3,1}}{t_{1,3}x + t_{2,3}y + t_{3,3}}, \frac{t_{1,2}x + t_{2,2}y + t_{3,2}}{t_{1,3}x + t_{2,3}y + t_{3,3}}, \right).$$

It will take a finite point off to infinity if the denominator vanishes. This will happen on the line whose coordinates are the last column of T.

10.6   A REMARK ABOUT SOLVING LINEAR SYSTEMS

A system of n linear equations in n unknowns can be written as a matrix equation

$$xA = b,$$

where A is an $n \times n$ matrix and has solution

$$x = bA^{-1}.$$

When A is 2×2 or 3×3, there is a formula for the inverse that is easy to program and relatively efficient. The same formula is valid in any number of dimensions, but for 4×4 matrices or larger it becomes impractical compared with other numerical methods such as those involving Gauss elimination.

This formula involves the determinant of several matrices. The determinant $\det(A)$ of a square matrix A is a number that can be calculated in a variety of ways. One way is directly from its definition, which is, for a general matrix, rather complicated.

For 2×2 matrices, the formula is quite simple. If

$$A = \begin{bmatrix} a_{1,1} & a_{1,2} \\ a_{2,1} & a_{2,2} \end{bmatrix},$$

then

$$\det(A) = a_{1,1}a_{2,2} - a_{1,2}a_{2,1}.$$

But if A has size $n \times n$, then its determinant is a sum of signed products

$$\pm a_{1,c_1} a_{2,c_2} \cdots a_{n,c_n},$$

where the n-tuple $(c_1, \ldots, c_n)$ ranges over all permutations of $(1, 2, \ldots, n)$, and so there are $n!$ in all – a potentially very large number. How is the sign determined? It is equal to the parity $\operatorname{sgn}(c)$ of the number of **inversions** of the permutation c, the pairs $c_i > c_j$ with $i < j$. So

$$\det A = \sum_c \operatorname{sgn}(c)\, a_{1,c_1} a_{2,c_2} \cdots a_{n,c_n},$$

where c ranges over all permutations of $[1, n]$.

For $n = 3$ we have the following parity assignments:

Permutation	Inversions	Parity
$(1, 2, 3)$	$\emptyset$	1
$(2, 3, 1)$	$(3, 1), (2, 1)$	1
$(3, 1, 2)$	$(3, 1), (3, 2)$	1
$(2, 1, 3)$	$(1, 2)$	-1
$(1, 3, 2)$	$(3, 2)$	-1
$(3, 2, 1)$	$(3, 1), (3, 2), (2, 1)$	-1

If

$$A = \begin{bmatrix} a_{1,1} & a_{1,2} & a_{1,3} \\ a_{2,1} & a_{2,2} & a_{2,3} \\ a_{3,1} & a_{3,2} & a_{3,3} \end{bmatrix},$$

then this assignment of signs tells us that its determinant is

$$D = a_{1,1}a_{2,2}a_{3,3} + a_{1,2}a_{2,3}a_{3,1} + a_{1,3}a_{2,1}a_{3,2} - a_{1,2}a_{2,1}a_{3,3} - a_{1,1}a_{2,3}a_{3,2}$$
$$- a_{1,3}a_{2,2}a_{3,1}.$$

The formula for this 3×3 determinant can be remembered easily if you note that the + terms are products going in a sense from upper left to lower right

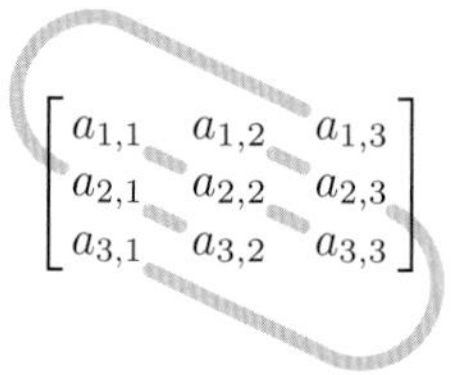

$$\begin{bmatrix} a_{1,1} & a_{1,2} & a_{1,3} \\ a_{2,1} & a_{2,2} & a_{2,3} \\ a_{3,1} & a_{3,2} & a_{3,3} \end{bmatrix}$$

and the $-$ terms from lower left to upper right

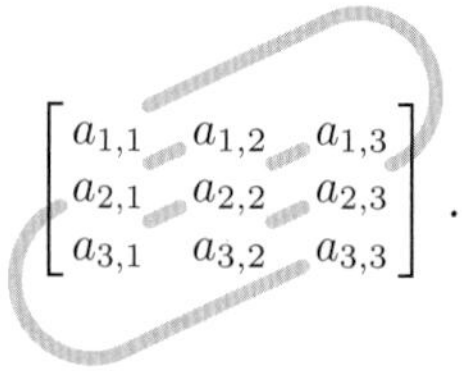

$$\begin{bmatrix} a_{1,1} & a_{1,2} & a_{1,3} \\ a_{2,1} & a_{2,2} & a_{2,3} \\ a_{3,1} & a_{3,2} & a_{3,3} \end{bmatrix}.$$

If A is any $n \times n$ matrix, its **cofactor matrix** A^*, sometimes called its **adjoint**, is a new $n \times n$ matrix derived from it. For any $1 \le i \le n$, $1 \le j \le n$ the associated (i, j)th **minor** is the determinant of the $(n-1) \times (n-1)$ matrix you get by cutting out from A the ith row and jth column. The cofactor matrix is that obtained from A whose (i, j)-entry $A^*_{i,j}$ is that minor multiplied by $(-1)^{i+j}$. For example, the following figure shows that the cofactor entry $A^*_{1,2}$ is $-(a_{2,1}a_{3,3} - a_{2,3}a_{3,1})$.

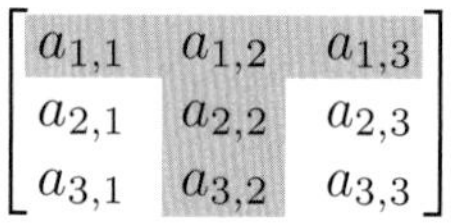

$$\begin{bmatrix} a_{1,1} & a_{1,2} & a_{1,3} \\ a_{2,1} & a_{2,2} & a_{2,3} \\ a_{3,1} & a_{3,2} & a_{3,3} \end{bmatrix}$$

The cofactor matrix is involved in another formula for the determinant:

$$\det(A) = \sum_{j=1}^{n} a_{i,j} A^*_{i,j}$$

for any choice of i. I'll leave it as an exercise to see that the two general formulas agree.

I recall also that the **transpose** ^tA of a matrix A is the matrix obtained by reflecting it around its principal diagonal. The formula for the determinant has as its immediate consequence that

■ *The inverse of a matrix A is the transpose of its cofactor matrix divided by its determinant.*

For example, if

$$A = \begin{bmatrix} a & b \\ c & d \end{bmatrix},$$

then its cofactor matrix is

$$A^* = \begin{bmatrix} d & -c \\ -b & a \end{bmatrix},$$

its transpose cofactor matrix is

$$^tA^* = \begin{bmatrix} d & -b \\ -c & a \end{bmatrix},$$

and its inverse is

$$A^{-1} = \frac{1}{ad - bc} \begin{bmatrix} d & -b \\ -c & a \end{bmatrix}.$$

You can check this if you multiply it by A.

The cofactor matrix of the 3×3 matrix $A = (a_{i,j})$ is

$$A^* = \begin{bmatrix} \begin{vmatrix} a_{2,2} & a_{2,3} \\ a_{3,2} & a_{3,3} \end{vmatrix} & -\begin{vmatrix} a_{2,1} & a_{2,3} \\ a_{3,1} & a_{3,3} \end{vmatrix} & \begin{vmatrix} a_{2,1} & a_{2,2} \\ a_{3,1} & a_{3,2} \end{vmatrix} \\ -\begin{vmatrix} a_{1,2} & a_{1,3} \\ a_{3,2} & a_{3,3} \end{vmatrix} & \begin{vmatrix} a_{1,1} & a_{1,3} \\ a_{3,1} & a_{3,3} \end{vmatrix} & -\begin{vmatrix} a_{1,1} & a_{1,2} \\ a_{3,1} & a_{3,2} \end{vmatrix} \\ \begin{vmatrix} a_{1,2} & a_{1,3} \\ a_{2,2} & a_{2,3} \end{vmatrix} & -\begin{vmatrix} a_{1,1} & a_{1,3} \\ a_{2,1} & a_{2,3} \end{vmatrix} & \begin{vmatrix} a_{1,1} & a_{1,2} \\ a_{2,1} & a_{2,2} \end{vmatrix} \end{bmatrix}.$$

Note that the signs form a checkerboard pattern like this:

$$\begin{bmatrix} + & - & + \\ - & + & - \\ + & - & + \end{bmatrix}.$$

You can check explicitly that $A\,{}^tA^* = I$.

EXERCISE 10.2. *Construct a PostScript procedure with one argument, a 3×3 matrix, that returns its inverse matrix. In this exercise, a 3×3 matrix should be an array of three arrays of three numbers, which are its rows:*

```
[[1 0 0][0 1 0][0 0 1]]
```

is the identity matrix.

10.7   THE GIMP PERSPECTIVE TOOL REVISITED

How does the theorem relate to the perspective tool? We start off with a rectangular image with corners at C_i for $i = 1, 2, 3, 4$. The mouse picks four points P_i somewhere in the image and calculates the perspective transformation, taking each C_i to P_i, or equivalently the 3×3 matrix, taking each

$$(C_{i,x}, C_{i,y}, 1) \longmapsto (P_{i,x}, P_{i,y}, 1)$$

in the sense of homogeneous coordinates. We can calculate this matrix explicitly. Let

$$\varepsilon_1 = (1, 0, 0)$$
$$\varepsilon_2 = (0, 1, 0)$$
$$\varepsilon_3 = (0, 0, 1)$$
$$\varepsilon_4 = (1, 1, 1).$$

Let σ be the 3×3 matrix taking the ε_i to the C_i, and τ the one taking them to the P_i. Then $\sigma^{-1}\tau$ takes the C_i to the P_i.

Finally, the perspective transform tool first shows the effect of this matrix on a grid and then, when it has been accepted, applies it to every pixel in the image. Incidentally, the matrix that the tool displays is that of the projective transformation we have constructed except that GIMP follows mathematical conventions rather than those of computer graphics, and so it is in fact the transpose of the one seen here.

EXERCISE 10.3. *Construct a PostScript program that reproduces with some flexibility the GIMP window at the beginning of this chapter. That is, it displays two 5×5 grids, one the transform of the other by a projective transformation corresponding to any given 3×3 matrix. Then set that matrix equal to the transpose of the one in the tool window,*

$$\begin{bmatrix} 0.535 & -0.219 & 39 \\ 0 & 0.5 & 24 \\ 0 & -0.00205 & 1 \end{bmatrix},$$

and verify that what you see is what you should see. There is one thing to be careful about – the origin of GIMP coordinates is at the upper left with y increasing downwards. So start your drawing by translating your origin to the upper left corner and then writing `1 -1 scale`*.*

10.8 PROJECTIONS IN 2D

One special case of a transformation nicely described by 3×3 homogeneous matrices is projection onto a line from a point. It is not a projective transformation because it collapses 2D onto 1D and is hence singular with no inverse. Suppose $P = (x_P, y_P)$ is the point from which things are projected and $f(x, y) = Ax + By + C = 0$ the line onto which things are projected. A point Q is transformed to the intersection of the line PQ with the line $f(x, y) = 0$ on the assumption it is not parallel to that line. This will be the point $R = (1 - t)P + tQ$, where $f(R) = 0$. We solve

$$f(R) = (1 - t)f(P) + tf(Q)$$
$$= 0$$
$$t(f(Q) - f(P)) = -f(P)$$
$$t = \frac{-f(Q)}{f(P) - f(Q)}$$
$$R = \frac{f(P)Q - f(Q)P}{f(P) - f(Q)}.$$

If we embed 2D in 3D, then P becomes $(x_P, y_P, 1)$, Q becomes $(x, y, 1)$, and

$$R = (f(P)Q - f(Q)P, \ f(P) - f(Q))$$
$$= f(P)(x, y, 1) - f(Q)(x_P, y_P, 1)$$
$$= f(P)Q - f(Q)P.$$

This already looks promising because we no longer have to contemplate dividing by 0; instead, when PQ is parallel to $f = 0$ the intersection will be a point at infinity. It becomes even more promising if we note that the expression $f(P)Q - f(Q)P$ is linear as a function of Q, and so the projection arises from a linear transformation in 3D or in other words a 3×3 matrix.

EXERCISE 10.4. *What is the matrix?*

10.9 PERSPECTIVE IN 3D

One way in which projective transformations originate is in viewing a plane in perspective in 3D. I'll look at just one example. Let's place an observer in 3D at the point $E = (0, 0, e)$ with $e > 0$ with the observer looking down the z-axis in the negative direction. We are going to have the observer looking at images by projecting them onto the plane $z = 0$. That is, we are going to map points

$P = (x, y, z)$ onto the point of intersection of the line from P to E with the viewing plane. This map is explicitly

$$(x, y, z) \longmapsto \left(\frac{ex}{e - z}, \frac{ey}{e - z} \right).$$

I'll now restrict this map to a plane of the form $y = y_0$, and so we have a 2D map

$$(x, z) \longmapsto \left(\frac{ex}{e - z}, \frac{ey_0}{e - z} \right).$$

I claim that this can be seen as a projective transformation – that is, calculated in terms of a 3×3 homogeneous matrix acting on homogeneous coordinates. This is not quite obvious. Later on when we look at 3D drawing in more detail, we'll see it as a special case of a very general fact, but it's worthwhile to do a simple case first.

First we change (x, z) into $(x, z, 1)$ and then into (x, z, w). The map above then becomes the restriction to $(x, z, 1)$ of the homogeneous linear map

$$(x, z, w) \longmapsto (ex, ey_0, w, ew - z)$$

or

$$[\, x \quad z \quad w \,] \longmapsto [\, x \quad z \quad w \,] \begin{bmatrix} e & & \\ & & -1 \\ ey_0 & & e \end{bmatrix}.$$

And of course these transforms may be combined with the techniques of Chapter 8.

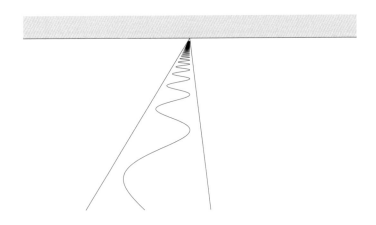

EXERCISE 10.5. *Construct a PostScript procedure with four 2D points P_i as arguments returning the 3×3 matrix by taking*

$$(0, 0) \mapsto P_0$$
$$(1, 0) \mapsto P_1$$
$$(0, 1) \mapsto P_2$$
$$(1, 1) \mapsto P_3.$$

REFERENCES

1. D. A. Brannan, M. Esplen, and J. Gray, **Geometry**, Cambridge University Press, 1999. The literature on projective geometry is large. With the advent of computers, the goals of the subject have become less abstract, and thus the older books tend seem somewhat out of date. This is a recent comprehensive text that illustrates the trend toward less abstraction. Lots of pictures.

2. M. Kemp, **The Science of Art: Optical Themes in Western Art from Brunelleschi to Seurat**, Yale University Pres, 1990. The literature on perspective is even more vast. This is an extremely thorough history with many implicit suggestions for projects in mathematical graphics.

3. Three books by the well-known expert in computer graphics, Jim Blinn,

    Jim Blinn's Corner: A Trip Down the Graphics Pipeline, Morgan Kaufmann, 1996
    Jim Blinn's Corner: Dirty Pixels, Morgan Kaufmann, 1998
    Jim Blinn's Corner: Notation, Notation, Notation, Morgan Kaufmann, 2003

    offer a well-motivated outsider's look at geometry. The essays are taken from Blinn's column in the journal *IEEE Computer Graphics and Applications*. Most are interesting, and some are absolutely delightful. Especially valuable are his essays on homogeneous coordinates, but there is much other mathematics and many mathematical pictures to be found in these volumes.

4. Andrew Glassner, **Andrew Glassner's Notebook: Recreational Computer Graphics**, Morgan Kaufmann, 1999. I mention this book here because it has much in common with Blinn's although not much with the topic of this chapter. Like Blinn's books, it is based on a column in *IEEE Computer Graphics and Applications* and amounts to a collection of improved essays from that column. It is more mathematical and less practical than Blinn's books. For that reason, not quite so interesting to a mathematician. Much fancier pictures!

5. The program **GIMP** is available on essentially all operating systems and without cost for Linux and Windows. Invaluable as an image manipulation tool

for most. Almost as capable as **Photoshop**®, and – for better or worse – about as easy to use. But for simple tasks it's fine. The **GIMP** home page is at

`http://www.gimp.org/`

6. Werner Boy, "Über die Curvatura integra," *Mathematische Annalen* (**57**), 1903, 151–184.
  This paper can be seen in the digitized edition of the Annalen at

  `http://134.76.163.65/agora_docs/25917TABLE_OF_CONTENTS.html`

  which is part of the Göttingen digitalization project.

7. Stephan Cohn-Vossen and David Hilbert, **Geometry and the Imagination**, Chelsea, 1952. Werner Boy was a student of Hilbert's, and his surface is said to have been a surprise to his advisor. Several variants of the surface are discussed on pages 317–321 of Hilbert and Cohn-Vossen but alas! without any history. Franois Apéry has written the beautiful book **Models of the Real Projective Plane** (Vieweg, 1987) on the same subject. It includes many handsome computer-generated pictures and explains how they were generated. Many other computer-generated pictures of the surface immersed in 3D can be found on the Internet, but I find Boy's original figures easiest to understand.

CHAPTER 11

Introduction to drawing in three dimensions

Drawing figures in 3D is considerably more complicated than drawing in 2D because we want to create the illusion of three dimensions on a two-dimensional page. There are in fact different ways to deal with this problem. I will illustrate a few ideas by showing a progression of pictures of a cube.

The first is a simple **orthogonal projection** of the **frame** of the cube. This just renders the image of the cube projected by parallel lines onto the (x, y) plane.

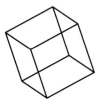

In the next we use **perspective projection**, which in effect renders the image of the cube by intersecting the (x, y) plane with rays from points in space to a fixed location, which plays the role of the eye. This scheme has the virtue that objects farther away from the eye look smaller. This provides an illusion of distance.

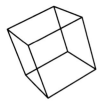

In the third we hide those faces of the cube that the eye lies behind. This adds the illusion of solidity.

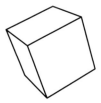

In the fourth we vary the darkness of the faces of the cube according to the degree to which they are oriented toward an imaginary vertical light source. We also get some color.

And in the fifth we embed the cube into a simple environment, even projecting a shadow from the same imaginary light source that affected brightness. Environment also adds visual cues for depth and orientation.

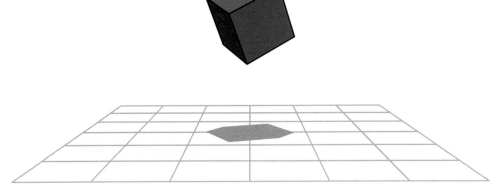

These effects are not too difficult to obtain with a 3D extension of PostScript. In the very next chapter we will see the mathematics on which they depend.

## CHAPTER 12

Transformations in 3D

There are no built-in routines for 3D drawing in PostScript. For this purpose we will need to use a library of PostScript procedures designed especially for the task as an extension to basic PostScript. In this chapter, we look at some of the mathematics behind such a library, which is much more complicated than that required in 2D.

We examine principally **rigid transformations**, those that affect an object without distorting it, but then at the end look at something related to shadow projections. The importance of rigid transformations is especially great because, to build an illusion of 3D through 2D images, the illusion of motion helps considerably. One point is that motion leads to an illusion of depth through size change and another is that it allows one to see an object from different sides. The motions used will be mostly rotations and translations, which are those that occur most commonly in the real world.

There are several reasons why it is a good idea to examine such transformations in dimensions one and two as well as three.

12.1 RIGID TRANSFORMATIONS

If we move an object around normally, it will not visibly distort – that is, to all appearances it will remain rigid. The points of the object themselves will move around, but the relative distances between points of the object will not change. We can formulate this notion more precisely. Suppose we move an object from one position to another. In this process, a point P will be moved to another point P_*. We will say that the points of the object are **transformed** into other points. A transformation is said to be **rigid** if it preserves relative distances – that is, if P and Q are transformed to P_* and Q_*, then the distance from P_* to Q_* is the same as that from P to Q.

We will take for granted something that can actually be proven but by a some-what distracting argument:

■ *A rigid transformation is* **affine**.

This means that if we have chosen a linear coordinate system in whatever context we are looking at (a line, a plane, or space), then the transformation $P \mapsto P_*$ is calculated in terms of coordinate arrays x and x_* according to the formula

$$x_* = xA + v,$$

where A is a matrix and v a vector. Another way of saying this is that first we apply a linear transformation whose matrix is A and then a translation by v. In 3D, for example, we require that

$$[\, x_* \ y_* \ z_* \,] = [\, x \ y \ z \,] A + [\, v_x \ v_y \ v_z \,].$$

The matrix A is called the **linear component**, and v is termed the **translation component** of the transformation.

It is clear that what we would intuitively call a rigid transformation preserves relative distances, but it might not be so clear that this requirement encapsulates rigidity completely. The following assertion may add conviction:

■ *A rigid transformation (in the sense I have defined it) preserves angles as well as distances.*

In other words, if P, Q, and R are three points transformed to P_*, Q_*, and R_*, then the angle θ_* between segments $P_* Q_*$ and $P_* R_*$ is the same as the angle θ between PQ and PR. This is because of the cosine law, which says that

$$\cos \theta_* = \frac{\|Q_* R_*\|^2 - \|P_* Q_*\|^2 - \|P_* R_*\|^2}{\|P_* Q_*\| \, \|P_* R_*\|}$$
$$= \frac{\|QR\|^2 - \|PQ\|^2 - \|PR\|^2}{\|PQ\| \, \|PR\|}$$
$$= \cos \theta.$$

A few other facts are more elementary:

■ *A transformation obtained by performing one rigid transformation followed by another rigid transformation is itself rigid.*

■ *The inverse of a rigid transformation is rigid.*

In the second statement, it is implicit that a rigid transformation has an inverse. This is easy to see. First of all, any affine transformation will be invertible if and only if its linear component is. But if a linear transformation does not have an inverse, then it must collapse at least one nonzero vector into the zero vector and it cannot be rigid.

EXERCISE 12.1. *Recall exactly why it is that a square matrix with determinant equal to zero must transform at least one nonzero vector into zero.*

An affine transformation is rigid if and only if its linear component is, for translation certainly doesn't affect relative distances. To classify rigid transformations, we must thus classify the linear ones. We'll do that in a later section 12.5 after some coordinate calculations.

EXERCISE 12.2. *The inverse of the transformation* $x \mapsto Ax + v$ *is also affine. What are its linear and translation components?*

12.2 DOT AND CROSS PRODUCTS

In this section I'll recall some basic facts about vector algebra.

■ Dot products

In any dimension the dot product of two vectors

$$u = (x_1, x_2, \ldots, x_n), \quad v = (y_1, y_2, \ldots, y_n)$$

is defined to be

$$u \bullet v = x_1 y_1 + x_2 y_2 + \cdots x_n y_n.$$

The relation between dot products and geometry is expressed by the cosine rule for triangles, which asserts that if θ is the angle between u and v, then

$$\cos \theta = \frac{u \bullet v}{\|u\| \, \|v\|}.$$

In particular, u and v are perpendicular when $u \bullet v = 0$.

■ Parallel projection

One important use of dot products and cross products will be in calculating various projections.

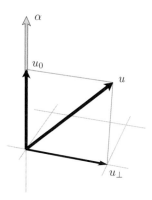

Suppose α to be any vector in space and u some other vector in space. The **projection of u along α** is the vector u_0 we get by projecting u perpendicularly onto the line through α.

How is this projection calculated? It must be a multiple of α. We can figure out which multiple by using trigonometry. We know three facts:

(a) The angle θ between α and u is determined by the formula

$$\cos\theta = \frac{u \bullet \alpha}{\|\alpha\| \, \|u\|}.$$

(b) The length of the vector u_0 is $\|u\| \cos\theta$, and this is to be interpreted algebraically in the sense that if u_0 faces in the direction opposite to α it is negative.

(c) The direction of u_0 is either the same or opposite to α. The vector $\alpha/\|\alpha\|$ is a vector of unit length pointing in the same direction as α. Therefore,

$$u_0 = \|u\| \cos\theta \, \frac{\alpha}{\|\alpha\|} = \|u\| \, \frac{u \bullet \alpha}{\|\alpha\| \, \|u\|} \, \frac{\alpha}{\|\alpha\|} = \left(\frac{u \bullet \alpha}{\|\alpha\|^2}\right)\alpha = \left(\frac{u \bullet \alpha}{\alpha \bullet \alpha}\right)\alpha.$$

■ Volumes

The basic result about volumes is that in any dimension n the **signed volume** of the parallelepiped spanned by vectors $v_1, v_2, \dots, v_n$ is the determinant of the matrix

$$\begin{bmatrix} v_1 \\ v_2 \\ \dots \\ v_n \end{bmatrix}$$

whose rows are the vectors v_i.

Let me illustrate this in dimension 2. First of all, what do I mean by the **signed** volume?

A pair of vectors u and v in 2D determine not only a parallelogram but an **orientation**, a sign. The sign is positive if u is rotated in a positive direction through the parallelogram to reach v and negative in the other case. If u and v happen to lie on the same line, then the parallelogram is degenerate and the sign is 0. The signed area of the parallelogram they span is this sign multiplied by the area of the parallelogram itself. Thus, in the figure on the left the sign is positive whereas on the right it is negative. Notice that the sign depends on the order in which we list u and v.

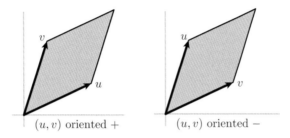

(u, v) oriented $+$ (u, v) oriented $-$

I'll offer two proofs that in 2D the signed area of the parallelogram spanned by two vectors u and v is the determinant of the matrix whose rows are u and v. The first is the simplest, but it has the drawback that it does not extend to higher dimensions although it does play an important role.

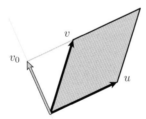

In the first argument, recall that $u_\perp$ is the vector obtained from u by rotating it positively through $90°$. If $u = [x, y]$, then $u_\perp = [-y, x]$. As the following figure illustrates, the signed area of the paralleogram spanned by u and v is the product of $\|u\|$ and the signed length of the projection of v onto the line through $u_\perp$.

Thus, the area is

$$\left(\frac{[-u_y, u_x] \bullet [v_x, v_y]}{\|u\|} \right) \|u\| = u_x v_y - u_y v_x = \det \begin{bmatrix} u_x & u_y \\ v_x & v_y \end{bmatrix} = \begin{vmatrix} u_x & u_y \\ v_x & v_y \end{vmatrix}.$$

The starting observation of the second argument is that shearing one of the rows of the matrix changes neither the determinant nor the area. For the area, this is because shears don't change area, whereas for the determinant it is a simple calculation:

$$\begin{vmatrix} x_1 + cx_2 & y_1 + cy_2 \\ x_2 & y_2 \end{vmatrix} = \begin{vmatrix} x_1 & y_1 \\ x_2 & y_2 \end{vmatrix}.$$

Thus, in each of the transitions below, neither the determinant nor the area changes. Because they agree in the final figure, where the matrix is a diagonal matrix, they must agree in the first.

There are exceptional circumstances in which one has to be somewhat fussy, but this is nonetheless the core of a valid proof that in dimension two determinants and area agree. It is still, with a little care, valid in dimension three. It is closely related to the Gauss elimination process.

EXERCISE 12.3. *Suppose A to be the matrix whose rows are the vectors v_1 and v_2. The argument above means that for most A we can write*

$$\begin{bmatrix} 1 & \\ c_2 & 1 \end{bmatrix} \begin{bmatrix} 1 & c_1 \\ & 1 \end{bmatrix} A = D,$$

where D is a diagonal matrix. Prove this, being careful about exceptions. For which A in 2D does it fail? Provide an explicit example of an A for which it fails. Show that even for such A (in 2D) the equality of determinant and area remains true.

In n dimensions, the Gaussian elimination process finds for any matrix A a permutation matrix w, a lower triangular matrix ℓ, and an upper triangulation matrix u such that

$$A = \ell u w.$$

The second argument in 2D shows that the claim is reduced to the special case of a permutation matrix, in which case it is clear.

■ Cross products

In 3D – and essentially only in 3D – there is a kind of product that multiplies two vectors to get another vector. If

$$u = (x_1, x_2, x_3), \quad v = (y_1, y_2, y_3),$$

then their **cross product** $u \times v$ is the vector

$$(x_2 y_3 - y_2 x_3, x_3 y_1 - x_1 y_3, x_1 y_2 - x_2 y_1).$$

This formula can be remembered if we write the vectors u and v in a 2×3 matrix

$$\begin{bmatrix} x_1 & x_2 & x_3 \\ y_1 & y_2 & y_3 \end{bmatrix}$$

and then for each column of this matrix calculate the determinant of the 2×2 matrix we get by crossing out in turn each of the columns. The only tricky part is that *with the middle coefficient we must reverse sign.* Thus

$$u \times v = \left(\begin{vmatrix} x_2 & x_3 \\ y_2 & y_3 \end{vmatrix}, -\begin{vmatrix} x_1 & x_3 \\ y_1 & y_3 \end{vmatrix}, \begin{vmatrix} x_1 & x_2 \\ y_1 & y_2 \end{vmatrix} \right).$$

It makes a difference in which order we write the terms in the cross product. More precisely

$$u \times v = -v \times u.$$

Recall that a vector is completely determined by its direction and its magnitude. The geometrical significance of the cross product is therefore contained in these two rules, which specify the cross product of two vectors uniquely.

■ *The length of $w = u \times v$ is the area of the parallelogram spanned in space by u and v.*
■ *The vector w lies in the line perpendicular to the plane containing u and v and its direction is determined by the right-hand rule: curl the fingers so as to go from u to v and the cross product will lie along your thumb.*

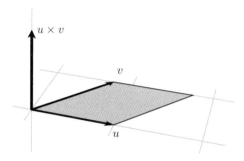

The cross product $u \times v$ will vanish only when the area of this parallelogram vanishes, when u and v lie in a single line, or equivalently, when they are multiples of one another.

That the length of the cross product is equal to the area of the parallelogram is a variant of Pythagoras's theorem applied to areas; it says precisely that the square of the area of a parallelogram is equal to the sum of the squares of the areas of its projections onto the coordinate planes. I do not know a really simple way to see that it is true, however. The simplest argument I know of is to start with the determinant formula for volume. If u, v, and w are three vectors, then on the one hand the volume of the parallelopiped they span is the determinant of the matrix,

$$\begin{bmatrix} u \\ v \\ w \end{bmatrix}$$

which can be written as the dot product of u and $v \times w$, whereas on the other hand this volume is the product of the projection of u onto the line perpendicular to the plane spanned by v and w and the area of the parallelogram spanned by v and w. A similar formula is valid in all dimensions, where it becomes part of the theory of **exterior products** of vector spaces, which is a topic beyond the scope of this book.

■ Perpendicular projection

Now let $u_\perp$ be the projection of u onto the plane perpendicular to α.

The vector u has the orthogonal decomposition

$$u = u_0 + u_\perp,$$

and therefore we can calculate

$$u_\perp = u - u_0.$$

Incidentally, in all of this discussion it is only the direction of α that plays a role. It is often useful to normalize α right at the beginning of these calculations, that is, to replace α by $\alpha/\|\alpha\|$ for instance.

Any good collection of PostScript procedures for 3D drawing will contain ones to calculate dot products, cross products, u_0 and $u_\perp$.

12.3 LINEAR TRANSFORMATIONS AND MATRICES

It is time to recall the precise relationship between linear transformations and matrices. The link is the notion of **frames**. A frame in any number of dimensions d is a set of d linearly independent vectors of dimension d. Thus, a frame in 1D is just a single nonzero vector. A frame in 2D is a pair of vectors whose directions do not lie in a single line. A frame in 3D is a set of three vectors not all lying in the same plane.

A frame in dimension d determines a coordinate system of dimension d and vice versa. If the frame is made up of $e_1, \ldots, e_d$, then every other vector x of dimension d can be expressed as a linear combination

$$x = x_1 e_1 + \cdots + x_d e_d,$$

and the coefficients x_i are its coordinates with respect to that basis. They give rise to the representation of x as a row array

$$[x_1 \ \ldots \ x_d].$$

Conversely, given a coordinate system, the vectors

$$e_1 = [1, 0, \ldots, 0], \quad \ldots, e_d = [0, 0, \ldots, 1]$$

are the frame giving rise to it.

I emphasize:

■ *A vector is a geometric entity with intrinsic significance – for example, the relative position of two points – or, if a unit of length has been chosen, something with direction and magnitude. In the presence of a frame, and only in the presence of a frame, it can be assigned coordinates.*

In other words, the vector is, if you will, an arrow, and it can be assigned coordinates only with respect to a given frame. Changing the frame changes the coordinates.

In a context where lengths are important, one usually works with **orthonormal frames** – those made up of a set of vectors e_i with each e_i of length 1 and distinct e_i and e_j perpendicular to each other:

$$e_i \bullet e_j = \begin{cases} 1 & \text{if } i = j \\ 0 & \text{otherwise.} \end{cases}$$

In the presence of any frame, vectors may be assigned coordinates. If the frame is orthonormal, the coordinates of a vector are given simply by

$$v = \sum c_i e_i, \quad c_i = e_i \bullet v.$$

If a particular point is fixed as origin, an arbitrary point may be assigned coordinates as well, namely, those of the vector by which the origin is displaced to reach that point.

Another geometric object is a **linear function**. A linear function ℓ assigns to every vector a number with the property, called **linearity**, that

$$\ell(au + bv) = a\ell(u) + b\ell(v).$$

A linear function ℓ may also be assigned coordinates, namely, the coefficients ℓ_i in its expression in terms of coordinates:

$$\ell(x) = \ell_1 x_1 + \cdots + \ell_d x_d.$$

A linear function is represented as a column array

$$\begin{bmatrix} \ell_1 \\ \cdots \\ \ell_d \end{bmatrix}.$$

Then $\ell(x)$ is the matrix product

$$[x_1 \ \cdots \ x_d] \begin{bmatrix} \ell_1 \\ \cdots \\ \ell_d \end{bmatrix}.$$

Another point to emphasize:

- *A linear function is a geometric entity with intrinsic significance that assigns a number to any vector. In the presence of a frame, and only in the presence of a frame, it can be assigned coordinates, the coefficients of its expression with respect to the coordinates of vectors.*

In physics, what I call vectors are called contravariant vectors, and linear functions are called covariant ones. There is often some confusion about these notions because often one has at hand an orthonormal frame and a notion of length, and given those each vector u determines a linear function f_u:

$$f_u(v) = u \bullet v.$$

A **linear transformation** T assigns vectors to vectors – also with a property of linearity. In the presence of a coordinate system, it, too, may be assigned coordinates, the array of coefficients $t_{i,j}$ such that

$$(xT)_j = \sum_{i=1}^{d} x_i t_{i,j}.$$

For example, in 2D we have

$$xT = [x_1 \ x_2] \begin{bmatrix} t_{1,1} & t_{1,2} \\ t_{2,1} & t_{2,2} \end{bmatrix}.$$

The rectangular array $t_{i,j}$ is called the linear **matrix** assigned to T in the presence of the coordinate system. In particular, *the ith row of the matrix is the coordinate array of the image of the frame element e_i with respect to T.* For example,

$$[1 \ 0] \begin{bmatrix} t_{1,1} & t_{1,2} \\ t_{2,1} & t_{2,2} \end{bmatrix} = [t_{1,1} \ t_{1,2}]$$

$$[0 \ 1] \begin{bmatrix} t_{1,1} & t_{1,2} \\ t_{2,1} & t_{2,2} \end{bmatrix} = [t_{2,1} \ t_{2,2}].$$

A final point, then, to emphasize is that

- *A linear transformation is a geometric entity with intrinsic significance transforming a vector into another vector. In the presence of a coordinate system (i.e., a frame), and only in the presence of a coordinate system, it can be assigned a matrix.*

In summary, *it is important to distinguish between intrinsic properties of a geometric object and properties of its coordinates.*

If A is any square matrix, one can calculate its **determinant**. It is perhaps surprising to know that this has intrinsic significance.

- *Suppose T to be a linear transformation in dimension d and A the matrix associated to T with respect to some coordinate system. The determinant of A is the factor by which T scales all d-dimensional volumes.*

This explains why, for example, the determinant of a matrix product AB is just the same as $\det(A)\det(B)$. If A changes volumes by a factor $\det(A)$ and B by the factor $\det(B)$, then the composite changes them by the factor $\det(A)\det(B)$.

If this determinant is 0, for example, it means that T is degenerate: it collapses d-dimensional objects down to something of lower dimension.

What is the significance of the sign of the determinant? Let's look at an example, where T amounts to 2D reflection in the y-axis. This takes $(1, 0)$ to $(-1, 0)$ and $(0, 1)$ to itself, and so the matrix is

$$\begin{bmatrix} -1 & \\ & 1 \end{bmatrix},$$

which has determinant -1. Here is its effect on a familiar shape:

$$R \longrightarrow Я$$

Now the transformed letter **R** is qualitatively different from the original – there is no continuous way to deform one into the other without some kind of one-dimensional degeneration. In effect, reflection in the y-axis changes **orientation** in the plane. This is always the case:

■ *A linear transformation with negative determinant changes orientation.*

12.4 CHANGING COORDINATE SYSTEMS

Vectors, and linear functions and linear transformations all acquire coordinates in the presence of a frame. What happens if we change frames?

The first question to be answered is, *How can we describe the relationship between two frames?*

The answer is, by a matrix. To see how this works, let's look at the 2D case first. Suppose e_1, e_2 and f_1, f_2 are two frames in 2D. Make up column matrices E and F whose entries are vectors instead of numbers – vectors chosen from the frames

$$e = \begin{bmatrix} e_1 \\ e_2 \end{bmatrix}, \quad f = \begin{bmatrix} f_1 \\ f_2 \end{bmatrix}.$$

We can write

$$f_1 = f_{1,1}e_1 + f_{1,2}e_2$$
$$f_2 = f_{2,1}e_1 + f_{2,2}e_2,$$

and it seems natural to define the 2×2 matrix

$$T_f^e = \begin{bmatrix} f_{1,1} & f_{1,2} \\ f_{2,1} & f_{2,2} \end{bmatrix}.$$

Then

$$f = T_f^e e,$$

and so the matrix T_f^e relates the two frames. This use of a matrix relating two different coordinate systems is conceptually quite different from that in which it describes a linear transformation.

In general, suppose e and f are two frames. Each of them is a set of d vectors. Let T_f^e be the matrix whose ith row is the array of coordinates of f_i with respect to e. We can write this relationship in an equation $f = T_f^e e$. This may seem a bit cryptic. In this equation, we think of the frames e and f themselves as column matrices

$$e = \begin{bmatrix} e_1 \\ \dots \\ e_d \end{bmatrix}, \quad f = \begin{bmatrix} f_1 \\ \dots \\ f_d \end{bmatrix}$$

whose entries are vectors, not numbers. One odd thing to keep in mind is that, when we apply a matrix to a vector we write the matrix on the right, but when we apply it to a frame we put it on the left.

Now suppose x to be a vector whose array of coordinates with respect to e is x_e. We can write this relationship as

$$x = x_e\, e = [x_1 \ \dots \ x_d] \begin{bmatrix} e_1 \\ \dots \\ e_d \end{bmatrix}$$

and again express the frame as a column of vectors.

With respect to the frame f we can write

$$x = x_f\, f.$$

We want to know how to compute x_e in terms of x_f and vice versa. But we can write

$$x = x_f\, f = x_f (T_f^e e) = (x_f\, T_f^e) e = x_e\, e,$$

which implies:

■ *If e and f are two frames with $f = T_f^e e$, then for any vector x*

$$x_e = x_f\, T_f^e.$$

A similar result is this:

■ *If e and f are two frames with $f = T_f^e e$, then for any linear function ℓ*

$$\ell_e = (T_f^e)^{-1}\ell_f.$$

This is what has to happen because the function evaluation

$$x_f \ell_f = x_e \ell_e = x_e\, T_f^e \ell_e$$

is intrinsic.

Finally, we deal with linear transformations. We get from the first frame e a matrix A_e associated to A and from the second a matrix A_f. *What is the relationship* *between the matrices A_e and A_f?* We start out by recalling that the meaning of A_e can be encapsulated in the formula

$$(x\,A)_e = x_e\,A_e$$

for any vector x. In other words, the matrix A_e calculates the coordinates of $x\,A$ with respect to the frame e. Similarly,

$$(x\,A)_f = x_f\,A_f,$$

and then we deduce

$$\begin{aligned}
x_e\,A_e &= (x\,A)_e \\
&= (x\,A)_f\,T_f^e \\
&= (x_f\,A_f)\,A_f^e \\
&= x_f(A_f\,T_f^e) \\
&= (x_e(T_f^e)^{-1})\,A_f\,T_f^e \\
&= x_e((T_f^e)^{-1}\,A_f\,T_f^e).
\end{aligned}$$

Hence,

■ *If e and f are two frames with $f = T_f^e\,e$, then for any linear transformation T*

$$T_e = (T_f^e)^{-1}\,A_f\,T_f^e.$$

This is an extremely important formula. For example, it allows us to see immediately that the determinant of a linear transformation, which is defined in terms of a matrix associated to it, is independent of the coordinate system giving rise to the matrix. That's because

$$A_e = (T_f^e)^{-1}\,A_f\,T_f^e, \qquad \begin{aligned}\det(A_e) &= \det((T_f^e)^{-1}\,A_f\,T_f^e) \\ &= \det((T_f^e)^{-1})\det(A_f)\det(T_f^e) = \det(A_f).\end{aligned}$$

Two matrices A and $T^{-1}AT$ are said to be **similar**. They are actually equivalent in the sense that they are the matrices of the same linear transformation but with respect to different coordinate systems.

12.5   RIGID LINEAR TRANSFORMATIONS

Let A be a $d \times d$ matrix representing a rigid linear transformation T with respect to the frame e:

$$A = \begin{bmatrix} a_{1,1} & a_{1,2} & \cdots & a_{1,d} \\ a_{2,1} & a_{2,2} & \cdots & a_{2,d} \\ & & \cdots & \\ a_{d,1} & a_{d,2} & \cdots & a_{d,d} \end{bmatrix}.$$

Then $e_i A$ is equal to the ith row of A. Therefore, the length of the ith row of A is also 1. The angle between e_i and e_j is $90°$ if $i \neq j$, and therefore the angle between the ith and jth rows of A is also $90°$ and the two rows must be perpendicular to each other. In other words, the rows of A must be an orthogonal frame. In fact,

■ *A linear transformation is rigid precisely when its rows make up an orthogonal frame.*

Any such matrix A is said to be **orthogonal**.

The **transpose** $^t A$ of a matrix A has as its rows the columns of A and vice versa. By definition of the matrix product $A^t A$, its entries are the various dot products of the rows of A with the columns of $^t A$. Therefore,

■ *A matrix A is orthogonal if and only if*

$$A^t A = I$$

or, equivalently, if and only if its transpose is the same as as its inverse.

If A and B are two $n \times n$ matrices, then

$$\det(AB) = \det(A)\det(B).$$

The determinant of A is the same as that of its transpose. If A is an orthogonal matrix, then

$$\det(I) = \det(A)\det(^t A) = \det(A)^2;$$

thus,

■ *The determinant of an orthogonal matrix is ± 1.*

If $\det(A) = 1$, A preserves orientation; otherwise, it reverses orientation. As we have already seen, there is a serious qualitative difference between the two types. If we start with an object in one position and move it continuously, then the transformation describing its motion will be a continuous family of rigid transformations. The linear component at the beginning is the identity matrix with determinant 1. Because the family varies continuously, the linear component can never change the sign of its determinant and must therefore always be orientation preserving. A way to change orientation would be to reflect the object, as if in a mirror.

12.6 ORTHOGONAL TRANSFORMATIONS IN 2D

In 2D the classification of orthogonal transformations is very simple. First of all, we can rotate an object through some angle θ (possibly $0°$).

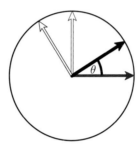

This preserves orientation. The matrix of this transformation is, as we saw in Chapter 2,

$$\begin{bmatrix} \cos\theta & \sin\theta \\ -\sin\theta & \cos\theta \end{bmatrix}.$$

Second, we can reflect things in an arbitrary line.

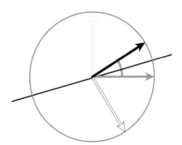

That is to say, given a line ℓ, we can transform points on ℓ into themselves and points in the line through the origin perpendicular to ℓ into their negatives. This reverses orientation.

EXERCISE 12.4. *If ℓ is the line at angle θ with respect to the positive x-axis, what is the matrix of this reflection?*

It turns out there are no more possibilities.

■ *Every linear rigid transformation in 2D is either a rotation or a reflection.*

Let $e_1 = [1, 0]$ and $e_2 = [0, 1]$, and let T be a linear rigid transformation. Since e_1 and e_2 both have length 1, both Te_1 and Te_1 also have length 1. All of these lie on the unit circle. Because the angle between e_1 and e_2 is $90°$, so is that between Te_1 and Te_2. There are two distinct possibilities, however. Either we rotate in the positive direction from Te_1 to Te_2 or in the negative direction.

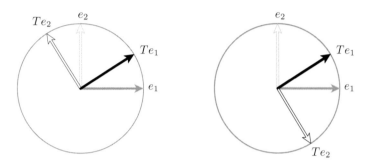

In the first case, we obtain Te_1 and Te_2 from e_1 and e_2 by a rotation. In the second case, something more complicated is going on. Here, as we move a vector u from e_1 towards e_2 and all the way around again to e_1, Tu moves along the arc from Te_1 to Te_2 all the way around again to Te_1 and *in the opposite direction*. Now if we start with two points anywhere on the unit circle and move them around in opposite directions, sooner or later they will meet. At that point we have $Tu = u$. Since T fixes u, it fixes the line through u and hence takes points on the line through the origin perpendicular to it into itself. It cannot fix the points on that line, and so it must negate them. In other words, T amounts to reflection in the line through u.

EXERCISE 12.5. *Explain why we can take u to be either of the points half way between e_1 and Te_1.*

EXERCISE 12.6. *Find a formula for the reflection of v in the line through u.*

12.7 ORTHOGONAL TRANSFORMATIONS IN 3D

There is one natural way to construct rigid linear motions in 3D. Choose an axis and choose on it a direction. Equivalently, choose a unit vector u and the axis to be the line through u with direction that of u.

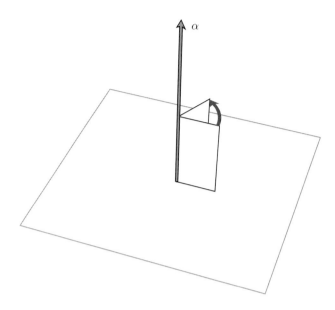

Choose an angle θ. Rotate things around the axis through angle θ in the positive direction as seen along the axis from the positive direction. This is called a **axial rotation**.

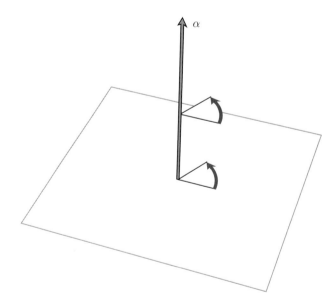

The motion can in some sense be decomposed into two parts. The plane through the origin perpendicular to the axis is rotated into itself, and points on the axis remain fixed. Therefore, the height of a point above the plane remains constant, and the projection of the motion onto this plane is just an ordinary 2D rotation.

■ *Every orientation-preserving linear rigid transformation in 3D is an axial rotation.*

I'll show this in one way here and in a slightly different way later on.

Recall that an **eigenvector** v for a linear transformation T is a nonzero vector v taken into a multiple of itself by T,

$$Tv = cv,$$

for some constant c, which is called the associated **eigenvalue**. This equation can be rewritten

$$Tv = cv = (T - cI)v = 0.$$

If $T - cI$ were invertible, then we would deduce from this that

$$v = (T - cI)^{-1}0 = 0,$$

which contradicts the assumption that $v \neq 0$. Therefore, $T - cI$ is not invertible, and $\det(T - cI) = 0$. In other words, c is a root of the **characteristic polynomial**

$$\det(A - xI),$$

where A is a matrix representing T and x is a variable. For a 3×3 matrix,

$$A - xI = \begin{bmatrix} a_{1,1} - x & a_{1,2} & a_{1,3} \\ a_{2,1} & a_{2,2} - x & a_{2,3} \\ a_{3,1} & a_{3,2} & a_{3,3} - x \end{bmatrix},$$

and the characteristic polynomial is a cubic polynomial that starts out

$$-x^3 + \cdots.$$

For $x < 0$ and $|x|$ large, this expression is positive, and for $x > 0$ and $|x|$ large it is negative. It must cross the x-axis somewhere, which means that it must have at least one real root. Therefore A has at least one real eigenvalue. In 2D this reasoning fails, for there may be two conjugate complex eigenvalues instead.

Let c be a real eigenvalue of T and v a corresponding eigenvector. Since T is a rigid transformation, $\|Tv\| = \|v\|$, or $\|cv\| = \|v\|$. Since $\|cv\| = |c|\|v\|$ and $\|v\| \neq 0$, $|c| = 1$ and $c = \pm 1$.

If $c = 1$, then we have a vector fixed by T. Since T preserves angles, it takes all vectors in the plane through the origin perpendicular to v into itself. Because T reserves orientation and $Tv = v$, the restriction of T on this plane also preserves orientation. Therefore, T rotates vectors in this plane and must be a rotation around the axis through v.

If $c = -1$, then we have $Tv = -v$. The transformation T still takes the complementary plane into itself. Since T preserves orientation in 3D but reverses orientation on the line through v, T reverses orientation on this plane. But then T must be a reflection on this plane according to the results of the previous section. We can find u such that $Tu = u$ and w perpendicular to u and v such that $Tw = -w$. In this case, T is rotation through $180°$ around the axis through u.

12.8   CALCULATING THE EFFECT OF AN AXIAL ROTATION

To begin this section, I remark again that to determine an axial rotation we must specify not only an axis and an angle but a direction on the axis. This is because the sign of a rotation in 3D is only determined if we know whether it is assigned by a left- or right-hand rule. At any rate, choosing a vector along an axis fixes a direction on it. Given a direction on an axis, I'll adopt the convention that the direction of positive rotation follows the right-hand rule.

So now the question we want to answer is this: *Consider a vector $\alpha \neq 0$ and an angle θ. If u is any vector in space and we rotate u around the axis through α by θ, what new point v do we get?* This is one of the basic calculations we will make to draw moved or moving objects in 3D.

There are some cases that are simple. If u lies on the axis, it is fixed by the rotation. If it lies on the plane perpendicular to α it is rotated by θ in that plane (with the direction of positive rotation determined by the right-hand rule).

If u is an arbitrary vector, we express it as a sum of two vectors, one along the axis and one perpendicular to it, and then use linearity to find the effect of the rotation on it.

To be precise, let R be the rotation we are considering. Given u we can find its **projection** onto the axis along α to be

$$u_0 = \left(\frac{\alpha \bullet u}{\alpha \bullet \alpha}\right) \alpha$$

Let $u_\perp$ be the projection of u perpendicular to α. It is equal to $u - u_0$. We write

$$u = u_0 + u_\perp$$
$$Ru = Ru_0 + Ru_\perp$$
$$= u_0 + Ru_\perp.$$

How can we find $Ru_\perp$?

Normalize α so $\|\alpha\| = 1$, in effect replacing α by $\alpha/\|\alpha\|$. This normalized vector has the same direction and axis as α. The vector $u_* = \alpha \times u_\perp$ will then be perpendicular to both α and to $u_\perp$ and will have the same length as $u_\perp$. The plane perpendicular to α is spanned by $u_\perp$ and u_*, which are perpendicular to each other and have the same length. The rotation R acts as a 2D rotation in the plane perpendicular to α, and so

■ *The rotation by θ takes $u_\perp$ to*

$$Ru_\perp = (\cos\theta)\,u_\perp + (\sin\theta)\,u_*.$$

In summary,

(1) Normalize α, replacing α by $\alpha/\|\alpha\|$.

(2) Calculate

$$u_0 = \left(\frac{\alpha \bullet u}{\alpha \bullet \alpha}\right)\alpha = (\alpha \bullet u)\,\alpha.$$

(3) Calculate

$$u_\perp = u - u_0.$$

(4) Calculate

$$u_* = \alpha \times u_\perp.$$

(5) Finally, set

$$Ru = u_0 + (\cos\theta)\,u_\perp + (\sin\theta)\,u_*.$$

EXERCISE 12.7. *What do we get if we rotate the vector $(1, 0, 0)$ around the axis through $(1, 1, 0)$ by $36°$?*

EXERCISE 12.8. *Write a PostScript procedure with α and θ as arguments that returns the matrix associated to rotation by θ around α.*

12.9 FINDING THE AXIS AND ANGLE

If we are given a matrix R that we know to be orthogonal and with determinant 1, how do we find its axis and rotation angle? As we have seen, it is a special case of the problem of finding eigenvalues and eigenvectors. But the situation is rather special and can be done in a more elementary manner. In stages,

(1) *How do we find its axis?* If e_i is the ith standard basis vector (one of **i**, **j**, or **k**) the ith column of R is Re_i. Now for any vector u the difference $Ru - u$ is

perpendicular to the rotation axis. Therefore, we can find the axis by calculating a cross product $(Re_i - e_i) \times (Re_j - e_j)$ for one of the three possible distinct pairs from the set of indices 1, 2, 3 – unless it happens that this cross product vanishes. Usually all three of these cross products will be nonzero vectors on the rotation axis, but in exceptional circumstances it can happen that one or more will vanish. It can even happen that all three vanish! But this only when A is the identity matrix, in which case we are dealing with the trivial rotation whose axis isn't well defined anyway.

At any rate, any of the three that is not zero will tell us what the axis is.

(2) *How do we find the rotation angle?*

As a result of part (1), we have a vector α on the rotation axis. Normalize it to have length 1. Choose one of the e_i so that α is not a multiple of e_i. Let $u = e_i$. Then Ru is the ith column of R.

Find the projection u_0 of u along α and set $u_\perp = u - u_0$. Calculate $Ru_\perp = Ru - u_0$. Next calculate

$$u_* = \alpha \times u_\perp$$

and let θ be the angle between $u_\perp$ and $Ru_\perp$. The rotation angle is θ if the dot product $Ru_\perp \bullet u_* \geq 0$; otherwise, $-\theta$.

EXERCISE 12.9. *If*

$$R = \begin{bmatrix} 0.899318 & -0.425548 & 0.100682 \\ 0.425548 & 0.798635 & -0.425548 \\ 0.100682 & 0.425548 & 0.899318 \end{bmatrix},$$

find the axis and angle.

12.10 EULER'S THEOREM

The fact that every orthogonal matrix with determinant 1 is an axial rotation may seem quite reasonable after some thought about what else such a linear transformation might be, but I don't think it is quite intuitive. To demonstrate this, let me point out that it implies that the combination of rotations around distinct axes is again a rotation. This is not at all obvious, and in particular it is difficult to see what the axis of the combination should be. This axis was constructed geometrically by Euler.

Let P_1 and P_2 be points on the unit sphere. Suppose P_1 to be on the axis of a rotation of angle θ_1 and P_2 on that of a rotation of angle θ_2. Draw the spherical arc from P_1 to P_2. On either side of this arc, at P_1 draw arcs making an angle of $\theta_1/2$ and at P_2 draw arcs making an angle of $\theta_2/2$. Let these side arcs intersect at α and β on the unit sphere. The rotation R_1 around P_1 rotates α to β, and the rotation R_2 around P_2 moves β back to α. Therefore, α is fixed by the composition $R_2 R_1$ and must be on its axis.

EXERCISE 12.10. *What is the axis of $R_1 R_2$ in the diagram above? Deduce from this result under what circumstances $R_1 R_2 = R_2 R_1$.*

12.11 MORE ABOUT PROJECTIONS

If P is a point in space and $f(x, y, z) = Ax + By + Cz + D = 0$ a plane, then just about any point other than P can be projected from P onto the plane. The formula for this is very simple. Suppose the point being projected is Q. The projection of Q onto the plane will be the point of the line through P and Q lying in the plane. The points of the line through P and Q are those of the form $R = (1 - t)P + tQ$, and so we must solve

$$f((1 - t)P + tQ) = (1 - t)f(P) + tf(Q) = 0$$

to get

$$R = \frac{f(P)\,Q - f(Q)\,P}{f(P) - f(Q)}.$$

The explicit formula for this is ugly unless we use 4D homogeneous coordinates. We embed 3D into 4D by setting the last coordinate 1, making

$$P = (a, b, c, 1), \quad Q = (x, y, z, 1),$$

and thus in homogeneous coordinates the projection formula becomes simply

$$R = f(P)Q - f(Q)P.$$

Projection in homogeneous coordinates is a linear transformation whose matrix is

$$f(P)I - \begin{bmatrix} A \\ B \\ C \\ D \end{bmatrix} [x_P \; y_P \; z_P \; w_P] = f(P)I - \begin{bmatrix} Ax_P & Ay_P & Az_P & Aw_P \\ Bx_P & By_P & Bz_P & Bw_P \\ Cx_P & Cy_P & Cz_P & Cw_P \\ Dx_P & Dy_P & Dz_P & Dw_P \end{bmatrix}.$$

This is an especially pleasant formula because it continues to make sense even if P is at infinity. Also, it respects our interpretation of points as row vectors and affine functions as column vectors.

## CHAPTER 13

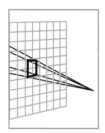

PostScript in 3D

This chapter explains a 3D extension to PostScript that I call `ps3d`, which you can find in the file `ps3d.inc`. To make this extension available, just put a line (ps3d.inc) run at the top of your program once you have downloaded `ps3d.inc`. The 3D graphics environment you get in this way is as close to the usual 2D one as I could make it with a few unusual features I'll mention later.

The underlying computations involved in 3D drawing are much more intensive than those for 2D, and in view of that it was surprising to me that package `ps3d` has turned out to be acceptably efficient. As I have mentioned already, I have used the usual PostScript routines in 2D as a model except that, instead of being restricted to affine transformations and therefore in this case to matrix arrays of size $3^2 + 3 = 12$, the underlying code works with arbitrary homogeneous 4×4 matrices or arrays of size 16. There are several reasons for doing this, and among them is that it makes the final perspective rendering simpler. This approach also makes it possible to cast shadows easily. But the principal reason, I have to confess, is mathematical simplicity. The disadvantage might be slowness, but although noticeable in some circumstances it doesn't seem to be a fatal problem. This whole package owes much to Jim Blinn's book **A Trip Down the Graphics Pipeline** mostly for the rigorous use of homogeneous coordinates throughout. On the other hand, using normal functions instead of just normal vectors is something only a mathematician would suggest happily. The advantage of doing this is that nonorthogonal transformations can be handled, although this is perhaps only a theoretical advantage because most coordinate changes are in fact orthogonal.

We continue to follow graphics conventions (i.e., as opposed to mathematical ones), and thus point vectors are rows and matrices are applied to them on the right. Affine functions $Ax + By + Cz + D$ will be expressed as column vectors.

Matrices multiply them on the left, and evaluation of such a function is a matrix product

$$
Ax + By + Cz + D = \begin{bmatrix} x & y & z & 1 \end{bmatrix} \begin{bmatrix} A \\ B \\ C \\ D \end{bmatrix}.
$$

I recall that composition on the stack is the principal motivation for these conventions of order in PostScript programming. Thus, $x\,S\,T$ conveniently applies S to x and then T. This conforms to the usual PostScript convention of applying operators to the object operated on so as to be ready to apply the next operator.

In this package, four dimensional directions are expressed in homogeneous coordinates as $[x, y, z, w]$. Such a point with $w \neq 0$ can be identified with the 3D point $(x/w, y/w, z/w)$, and one with $w = 0$ can be identified with a nonoriented direction in 3D. There is some inconsistency involved in this last point because sometimes the package implicitly uses oriented directions.

13.1 A SURVEY OF THE PACKAGE

In beginning 3D drawing, after loading the file ps3d.inc, the programmer has the option of choosing an eye position which, by convention, lies somewhere on the positive z-axis. The programmer is then looking down towards the negative z-axis, and what is seem is the projection of objects on the viewing screen, which is the plane, $z = 0$. On this plane, x and y are the usual 2D user coordinates. This projection of a point onto the viewing screen is the intersection of the line from that point to the observer's location with the plane $z = 0$. It is expected that most objects drawn will be located with $z < 0$.

If the programmer does not choose an eye location, the default mode of projection is orthogonal projection onto the plane $z = 0$, and thus the eye is effectively at ∞ on the z-axis, the 4D point $[0, 0, 1, 0]$. The point of the convention for looking down (rather than up) the z-axis is to get the 3D orientation to be that of the right-hand rule as well as to have the usual 2D coordinates interpreted naturally. Recall that, according to the right-hand rule, if the right fist is curled so as to indicate how the positive x-axis is to be rotated towards the positive y-axis, then the extended thumb points towards positive values of z.

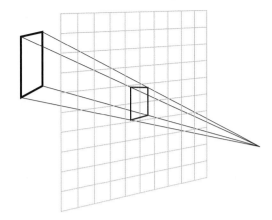

This setup has some paradoxical features. The following figures are identical except for the location of the eye, which is at $z = 27$, $z = 9$, and $z = 3$, respectively:

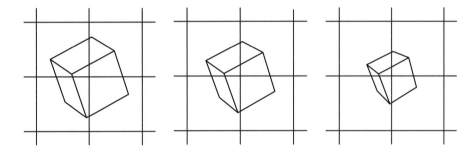

What's odd here is that as we move closer the cube becomes smaller! The explanation for this is that moving the eye in this model is not like moving your head back to change your view. When you move your head you move the whole eye, which includes both the lens at the front and the retina at the back. The distance from the lens to the retina – the **focal length** of the eye – does not change. Whereas in the graphics model I am using, moving the eye really means changing the focal length as well as the location of the eye. We are looking at the plane $z = 0$ as though it were a window onto which the world in the region $z \leq 0$ is projected, but as we move away our focal length is increased, too. In effect the left picture is taken with a telephoto lens, the right-hand one with a wide angle lens. The two changes, location and focal length, compensate for each other, and thus the grid, which is located on the plane $z = 0$, remains constant. This behavior might take getting used to. In the current version, the eye location is set once at the very beginning of a program and should never be changed again. In some future version I'll make it possible to move the eye around like a camera, changing location and also zooming in and out.

There is another odd feature of these conventions. As I have already mentioned, normally the 3D objects drawn will lie in the region $z < 0$ beyond the viewing screen.

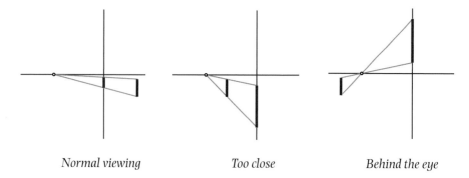

Normal viewing   *Too close*   *Behind the eye*

If the object lies between the eye and $z = 0$, it will be enlarged. If it lies behind the eye, it will be placed in an odd location. It is because of this unfamiliar behavior that it is best to draw only objects behind the plane $z = 0$.

The most frequently used commands in `ps3d.inc` are those that construct 3D paths – `moveto3d`, `lineto3d`, `closepath3d`, and their relative versions. There are also 3D coordinate changes such as `translate3d`, `scale3d`, and `rotate3d`. Rotations require an axis as well as an angle as argument. As with PostScript in 2D, there are implicitly two 3D coordinate systems. One is the user's 3D coordinate system, and the other is the default 3D system, which is also the user's system when 3D programming starts. Part of the 3D graphics environment keeps track of the transformation from the first to the second by means of a 4×4 matrix. Thus, when a program executes a command like `0 0 0 moveto3d`, this transformation is applied to the 3D point $(0, 0, 0)$ to get its default 3D coordinates, and then the projection of this point upon the viewing plane is calculated. In this way, a 2D path is constructed as the 3D drawing proceeds. The transformation from user 3D to default 3D coordinates is affected by the 3D coordinate changes. As in 2D, coordinate changes directly affect the frame with respect to which the user coordinates are interpreted. Thus,

```
[0 0 3 1] set-eye

[0 1 0] -45 rotate3d

newpath
-0.5 -0.5 0 moveto3d
0.5 -0.5 0 lineto3d
```

```
 0.5 0.5 0 lineto3d
-0.5 0.5 0 lineto3d
-0.5 -0.5 0 lineto3d
closepath
gsave
0.7 0 0 setrgbcolor
fill
grestore

stroke
```

Just as in ordinary 2D PostScript we use a graphics stack to manage 3D graphics environments. The 3D graphics environment at any moment amounts to two homogeneous linear transformations T (4×4) and D (4×3), the first transforming user 3D coordinates to the default ones and the second determining the collapse from 3D onto the viewing screen. The graphics stack is an array gstack3d of some arbitrary height gmax; it happens to be 64 in ps3d. But the part of the stack that's used bounces around. The current height of the stack at any moment is ght, and the current graphics data T and D are in gstack3d[ght]. The user will ordinarily not need to access the 3D graphics stack directly because there are operators that access its components directly. The 3D graphics stack is managed with the commands gsave3d and grestore3d. One use for the 3D graphics stack is to move different components of a figure independently or even in linkage with other components. (This technique is explained quite nicely in Chapter 3 of Blinn's book.)

The 3D graphics environment really requires only T and D but also, for efficiency and convenience, the inverse matrix T^{-1} is part of the stored data. Thus, an item on the 3D graphics stack is an array of three matrices: T, T^{-1}, and D. The first component is called the **current 3D transformation matrix**. The effect of transforming coordinates by a matrix R (rotation, translation, scaling, projection, etc.) is to replace T by RT and T^{-1} by $T^{-1}R^{-1}$. The third component D is called the **display matrix** and is used to transform homogeneous 4-vectors to homogeneous 3-vectors. At the moment the only display matrices used are those implementing perspective onto the plane $z = 0$ to an eye located in homogeneous space. Also at the moment it is set once and for all at the beginning of 3D drawing.

We will see why we want quick access to T^{-1} Section 13.5.

In Blinn's terms, 3D graphics involves a pipeline: a composition of coordinate changes from user coordinates to default 3D coordinates, then to the display plane (here $z = 0$), and finally to the page. The last step is controlled by the usual 2D PostScript commands. The first two are managed in this package.

Loading ps3d (with the command (ps3d.inc) run) causes a file matrix.inc of 3D matrix procedures to be loaded as well. It also causes the variables gstack3d, gmax, ght that determine the 3D graphics environment to be defined. Initially, ght is 0, T and T^{-1} are the 4×4 identity matrices, and D amounts to orthogonal projection along the positive z-axis.

In addition ps3d defines variables cpt3d and lm3d, the **current point** and the **last point moved to**, and 3D points expressed in default 3D coordinates. Normally, the user will have no need to know them.

The usual 3D vectors are identified with a 4D vector whose last coordinate is 1. Next I'll explain the commands grouped by subject.

13.2   THE 3D GRAPHICS ENVIRONMENT

	Arguments	*Command*	*Return value*
■	—	gsave3d	—

Puts a new copy of the current 3D environment on the graphics stack. It is extremely important to realize that manipulating the 3D graphics stack has no effect whatsoever on the 2D graphics state. And vice versa.

■	—	grestore3d	—

Restores the previous 3D graphics state.

■	—	cgfx3d	$[T\ U\ D]$
■	—	ctm3d	T
■	—	cim3d	T^{-1}
■	—	cdm3d	D
■	—	currentpoint3d	$[x\,y\,z]$

This returns the current point in user space T^{-1} applied to cpt3d.

■	—	display-matrix	D

Returns a 4×3 homogeneous matrix.

■	$[x\,y\,z\,w]$	set-eye	—

Sets the eye at (x, y, z, w). In practice, currently it is assumed that $x = y = 0$. Here, as elsewhere in 3D drawing, $w = 0$ indicates a direction in space, whereas $(x, y, z, 1)$ indicates an ordinary 3D point.

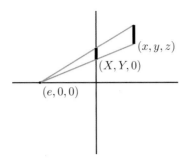

The 3D to 2D projection takes a point (x, y, z) to the intersection (X, Y) of the line from that point to the eye with the plane $z = 0$. Assuming the eye is at (e_x, e_y, e_z), we can find it explicitly as

$$(1 - t)(e_x, e_y, e_z) + t(x, y, z),$$

where $(1 - t)e_z + tz = 0$. Therefore,

$$t = \frac{e_z}{e_z - z}$$

$$1 - t = \frac{-z}{e_z - z}$$

$$X = \frac{xe_z - ze_x}{e_z - z}, \quad Y = \frac{ye_z - ze_y}{e - z}.$$

If the eye is at $(0, 0, e)$, this becomes

$$X = \frac{ex}{e - z}, \quad Y = \frac{ey}{e - z}.$$

In terms of homogeneous coordinates, it takes (x, y, z, w) to $(xe_z - ze_x, ye_z - ze_y, e_zw - ze_w)$. Equivalently, it sets the display matrix to be

$$\begin{bmatrix} e_z & 0 & 0 \\ 0 & e_z & 0 \\ -e_x & -e_y & -e_w \\ 0 & 0 & e_z \end{bmatrix}$$

| ■ | — | get-eye | $[x\,y\,z\,w]$ |
| ■ | — | get-virtual-eye | $[x\,y\,z\,w]$ |

This last will be explained later.

| ■ | $[x\,y\,z\,w]$ | render | $[X/W\ Y/W]$ |

This applies the display matrix to the original point and then turns the result $[X\ Y\ W]$ into a real 2D point in 2D user coordinates.

| ■ | $x\,y\,z$ | transform2d | $x'\ y'$ |

Transforms from user 3D coordinates first to default 3D coordinates and then to 2D user coordinates.

13.3 COORDINATE TRANSFORMATIONS

A coordinate transformation multiplies the current transformation matrix by the matrix of the transformation. The current matrix is an array of sixteen numbers, interpreted as

$$
\begin{bmatrix}
T_0 & T_4 & T_8 & T_{12} \\
T_1 & T_5 & T_9 & T_{13} \\
T_2 & T_6 & T_{10} & T_{14} \\
T_3 & T_7 & T_{11} & T_{15}
\end{bmatrix}.
$$

Let U be the inverse of T.

■ $x\,y\,z$ translate3d   —

The new current matrix becomes

$$
\begin{bmatrix}
1 & 0 & 0 & 0 \\
0 & 1 & 0 & 0 \\
0 & 0 & 1 & 0 \\
x & y & z & 1
\end{bmatrix}
\begin{bmatrix}
T_0 & T_4 & T_8 & T_{12} \\
T_1 & T_5 & T_9 & T_{13} \\
T_2 & T_6 & T_{10} & T_{14} \\
T_3 & T_7 & T_{11} & T_{15}
\end{bmatrix},
$$

and the inverse transformation becomes

$$
\begin{bmatrix}
U_0 & U_4 & U_8 & U_{12} \\
U_1 & U_5 & U_9 & U_{13} \\
U_2 & U_6 & U_{10} & U_{14} \\
U_3 & U_7 & U_{11} & U_{15}
\end{bmatrix}
\begin{bmatrix}
1 & 0 & 0 & 0 \\
0 & 1 & 0 & 0 \\
0 & 0 & 1 & 0 \\
-x & -y & -z & 1
\end{bmatrix}.
$$

■ $[x\,y\,z]\,A$ rotate3d   —

The arguments are axis and angle. The new current transformation matrix becomes

$$
\begin{bmatrix}
R_0 & R_3 & R_6 & 0 \\
R_1 & R_4 & R_7 & 0 \\
R_2 & R_5 & R_8 & 0 \\
0 & 0 & 0 & 1
\end{bmatrix}
\begin{bmatrix}
T_0 & T_4 & T_8 & T_{12} \\
T_1 & T_5 & T_9 & T_{13} \\
T_2 & T_6 & T_{10} & T_{14} \\
T_3 & T_7 & T_{11} & T_{15}
\end{bmatrix},
$$

and the inverse transformation is

$$
\begin{bmatrix}
U_0 & U_4 & U_8 & U_{12} \\
U_1 & U_5 & U_9 & U_{13} \\
U_2 & U_6 & U_{10} & U_{14} \\
U_3 & U_7 & U_{11} & U_{15}
\end{bmatrix}
\begin{bmatrix}
R_0 & R_1 & R_2 & 0 \\
R_3 & R_4 & R_5 & 0 \\
R_6 & R_7 & R_8 & 0 \\
0 & 0 & 0 & 1
\end{bmatrix}.
$$

since the inverse of a rotation matrix is its transpose.

■ $a\,b\,c$ scale3d   —

The new 3D transformation matrix is

$$
\begin{bmatrix} a & 0 & 0 & 0 \\ 0 & b & 0 & 0 \\ 0 & 0 & c & 0 \\ 0 & 0 & 0 & 1 \end{bmatrix}
\begin{bmatrix} T_0 & T_4 & T_8 & T_{12} \\ T_1 & T_5 & T_9 & T_{13} \\ T_2 & T_6 & T_{10} & T_{14} \\ T_3 & T_7 & T_{11} & T_{15} \end{bmatrix},
$$

and the inverse transformation is

$$
\begin{bmatrix} U_0 & U_4 & U_8 & U_{12} \\ U_1 & U_5 & U_9 & U_{13} \\ U_2 & U_6 & U_{10} & U_{14} \\ U_3 & U_7 & U_{11} & U_{15} \end{bmatrix}
\begin{bmatrix} a^{-1} & 0 & 0 & 0 \\ 0 & b^{-1} & 0 & 0 \\ 0 & 0 & c^{-1} & 0 \\ 0 & 0 & 0 & 1 \end{bmatrix}.
$$

■   matrix   concat3d   —

The argument is a 3×3 matrix M, that is, an array of nine numbers. This replaces T by M (extended) T.

The new 3D transformation matrix is

$$
\begin{bmatrix} M_0 & M_3 & M_6 & 0 \\ M_1 & M_4 & M_7 & 0 \\ M_2 & M_5 & M_8 & 0 \\ 0 & 0 & 0 & 1 \end{bmatrix}
\begin{bmatrix} T_0 & T_4 & T_8 & T_{12} \\ T_1 & T_5 & T_9 & T_{13} \\ T_2 & T_6 & T_{10} & T_{14} \\ T_3 & T_7 & T_{11} & T_{15} \end{bmatrix},
$$

and the inverse transformation is

$$
\begin{bmatrix} U_0 & U_4 & U_8 & U_{12} \\ U_1 & U_5 & U_9 & U_{13} \\ U_2 & U_6 & U_{10} & U_{14} \\ U_3 & U_7 & U_{11} & U_{15} \end{bmatrix}
\begin{bmatrix} M_0' & M_3' & M_6' & 0 \\ M_1' & M_4' & M_7' & 0 \\ M_2' & M_5' & M_8' & 0 \\ 0 & 0 & 0 & 1 \end{bmatrix}.
$$

■ $[A\,B\,C\,D]\,P$ plane-project   —

Projects from the 3D homogeneous point P onto a plane. After this change, which is not invertible, paths are projected onto the plane. Explicitly, the map here is $Q \mapsto f(P)\,Q - f(Q)\,P$.

The matrix transformation is

$$
\begin{bmatrix} f(P) - AP_0 & -AP_1 & -AP_2 & -AP_3 \\ -BP_0 & f(P) - BP_1 & -BP_2 & -BP_3 \\ -CP_0 & -CP_1 & f(P) - CP_2 & -CP_3 \\ -DP_0 & -DP_1 & -DP_2 & f(P) - DP_3 \end{bmatrix}
\begin{bmatrix} T_0 & T_4 & T_8 & T_{12} \\ T_1 & T_5 & T_9 & T_{13} \\ T_2 & T_6 & T_{10} & T_{14} \\ T_3 & T_7 & T_{11} & T_{15} \end{bmatrix}.
$$

The inverse transformation is invalid.

- $v = [x\,y\,z\,w]\,M$ transform3d $v\,M$
- $M\,f$ dual-transform3d $M\!f$

13.4 DRAWING

- $x\,y\,z$ moveto3d   —
- $x\,y\,z$ lineto3d   —
- $x\,y\,z$ rmoveto3d   —
- $x\,y\,z$ rlineto3d   —
- $x_1\,y_1\,z_1\ldots$ curveto3d   —

The input is nine numbers.

- $x_1\,y_1\,z_1\ldots$ rcurveto3d   —
- —   closepath3d   —

It updates the current 3D point.

- [see below]   mkpath3d   —

The arguments are t_0, t_1, N, an array of parameters, and the function /f. This function has two arguments: the array of parameters and a single variable t. It returns an array made up of (1) a 3D point (x, y, z) and (2) a velocity vector x', y', z'). This procedure draws the path in N pieces using f as parametrization.

- —   2d-path-convert   —

This converts the current 2D path to a 3D path according to the current 3D environment. Thus,

```
[0 0 5 1] set-eye

[0 1 0] 45 rotate3d

newpath
0 0 moveto
(ABC) true charpath
2d-path-convert
gsave
0.7 0 0.1 setrgbcolor
fill
grestore

stroke
```

ABC

13.5 SURFACES

Surface rendering is treated in much more detail in the next chapter, but I shall make a few preliminary remarks here.

Surfaces are most conveniently interpreted as an assembly of flat plates – maybe very small ones to be sure. To ensure an illusion of depth in 3D pictures, they are often shaded according to the position of some imaginary light source. In mathematical graphics we are not interested in realistic rendering, which may even clutter up a good diagram with irrelevant junk, and can settle for simple tricks. Still, a few things are important for the right illusion.

■ $x \, [a_0 \, a_1]$ 1shade   —
■ $x \, [a_0 \, a_1 \, a_2 \, a_3]$ shade   —
■ $x \, [a_0 \ldots a_n]$ bshade   —

The number x lies in $[-1, 1]$ and is usually the result of the calculation of the dot product of normal vector and light direction. It the number is shifted to $[0, 1]$, and then the weighting function is applied.

The coordinates a_0 and so on are control values for a Bernstein polynomial determining the shading weight. The values of a_0 and a_n are minimum and maximum. The first two are duplicated by the last one, but they are significantly more efficient. For shade itself, $[0 \; 1/3 \; 2/3 \; 1]$ would be neutral and $[0 \; 0 \; 1 \; 1]$ would be relatively strong contrast. These things are discussed in detail in the next chapter.

■   array of 3D points   normal-function $[A, B, C, D]$

Returns the normal function $Ax + By + Cz + D$, which is 0 on the points of the array and increasing in the direction determined by the right-hand rule.

It is in dealing with matters affecting the appearance of surfaces – visibility and shading – that we need T^{-1}. The point x is outside the piece of surface Σ if and only if $x \cdot f_\Sigma \geq 0$, where f_Σ is the normal function associated to S. Similarly, we often want to know if x is outside the transformed surface ST. This happens if and only if xT^{-1} is outside S or

$$xT^{-1} \cdot f_\Sigma = x \cdot T^{-1} f_\Sigma \geq 0,$$

from which we see that $f_{\Sigma T} = T^{-1} f_\Sigma$. We could calculate $T^{-1} f_\Sigma$ for each Σ, but in practice we will want to test "visibility" for many bits of surface Σ and only one x (the "eye," for instance). So we must be ready to calculate $T^{-1}x$. This and similar calculations involving the light source motivate the extra matrix T^{-1} on the graphics stack. The command get-virtual-eye returns this point.

13.6 CODE

The `ps3d` package is in `ps3d.inc`. This incorporates a collection of generic matrix routines imported from `matrix.inc`. This package is documented separately in `matrix.pdf`. Sample 3D drawing is contained in the files `cube-frame.ps`, `cube-solid.ps`, `cube-shaded.ps`, and `cube-shadow.ps`. The last two of these use techniques discussed in the next chapter.

REFERENCES

1. Modern perspective was discovered in the early fifteenth century, and early accounts are still of interest. Leon Alberti wrote the first treatise on it in 1435, and what he said cannot be bettered for succinctness: ". . . whoever looks at a picture sees a certain cross section of a visual pyramid." (This is quoted from p. 209 of **A Documentary History of Art**, edited and translated by Elizabeth Holt, published by Doubleday Anchor Books, 1957.)

2. Jim Blinn, **Jim Blinn's Corner – A Trip Down the Graphics Pipeline**, Morgan Kaufmann, San Francisco, 1996. Chapter 8 of this book examines interesting questions about allowing not only a change in 3D coordinates but also ways in which the display matrix can be changed to conform to different ways of viewing 3D space. In mathematical graphics, this isn't so important as it is in nonmathematical graphics, where one often has to track a moving object.

CHAPTER 14

Drawing surfaces in 3D

Only in mathematics books do spheres look like the thing on the left below rather than the one on the right.

What your eye sees in reality are fragments of surfaces or rather the light reflected from them. In computer graphics, a surface is an assembly of **flat plates**, each of which is a 2D polygon moved into location in space together with a specification of one of its two sides. One difference between surfaces drawn by computer and those in the real world is that in the real world surfaces possess detail down to microscopic size, including the appearance of smooth curvature. Surfaces drawn by a computer can only be an approximation of these. Sometimes the plates making up a surface will have some extra data added to them to help make the illusion of reality stronger.

PostScript is not efficient enough to do very realistic 3D rendering. Among other things, it does not have access to specialized 3D hardware and in particular has little comprehension of depth. But it is efficient enough to do a reasonable job on mathematical images.

14.1   FACES

A **surface fragment** or sometimes **face** will be an oriented polygon made up of 3D points all lying in some single plane. The orientation means a choice of **side** to the

217

polygon – a top as opposed to a bottom, or an outside as opposed to an inside. The orientation in practice means that the vertices of the polygon are arranged in an array going around the edges of the polygon according to the right-hand rule – so that if the right hand curls around in the direction the vertices are numbered, the thumb points toward the side chosen. In other words, if the polygon is on a plane in front of you and the vertices are arranged in a counterclockwise direction, your eye is on the outside.

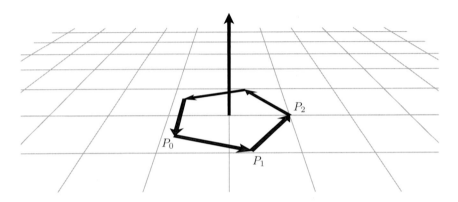

Associated to an oriented polygon in 3D is what I call its **normal function**, the unique linear function $Ax + By + Cz + D$ with these three properties:

- All points in the polygon lie on the plane $Ax + By + Cz + D = 0$;
- the outside of the face is where $Ax + By + Cz + D$ is positive;
- the length of the vector $[A, B, C]$ is 1.

The normal function can be computed from the oriented polygon each time it is to be drawn, but it is more efficient in designing a face to be drawn to compute the normal function once when the array is determined, and then carry it along as part of the structure of the face. Suppose the vertices of the face are P_0, P_1, P_2, ... We will always assume that the edges of a face are **nondegenerate** in the sense that none of its edges have length 0 and that no two successive edges lie in a straight line. In practice this sometimes takes a little work to guarantee, as we will see later on. At any rate, under this assumption the vector cross product

$$(P_1 - P_0) \times (P_2 - P_1)$$

will be nonzero and perpendicular to the polygon facing outwards. Let $[A, B, C]$ be this vector normalized by dividing it by its length. The value of $Ax + By + Cz$ will be the same on all vertices of the polygon, and if we set D to be the negative of this common value, then $Ax + By + Cz + D$ will be the normal function.

In the simplest of our methods of drawing surfaces, a face will be by convention an array of two other arrays, the first being the oriented array of points making up the oriented polygon and the second an array of four numbers making up the normal function. Thus,

```
[[[0 0 0] [1 0 0] [1 1 0] [0 1 0]] [0 0 1 0]]
```

is the face representing a unit square in the (x, y) plane facing out along the positive z-axis, whose normal function is z.

In the code in ps3d there is a procedure normal-function with one argument, an oriented array of 3D points, which returns the array [A B C D] corresponding to its normal function. It implements exactly the calculation described above. If the cross product is 0, the procedure returns an empty array.

14.2 POLYHEDRA

A **polyhedron** in 3D is a collection of (flat) faces, each of which is a surface fragment whose orientation points outward, making up the boundary of a 3D region with inside and outside. Cubes, for examples, are polyhedra. According to our convention, a polyhedron will be an array of faces, where a face is what is prescribed in the previous section. The following, for example, defines a complete unit cube, centered at $(1/2, 1/2, 1/2)$ with edges aligned parallel to the coordinate axes:

```
/cube [
   [[[0 0 1] [1 0 1] [1 1 1] [0 1 1]] [0 0 1 -1]]
   [[[0 1 0] [1 1 0] [1 0 0] [0 0 0]] [0 0 -1 0]]
   [[[0 0 0] [0 0 1] [0 1 1] [0 1 0]] [0 -1 0 0]]
   [[[1 1 0] [1 1 1] [1 0 1] [1 0 0]] [1 0 0 -1]]
   [[[0 1 0] [0 1 1] [1 1 1] [1 1 0]] [0 1 0 -1]]
   [[[1 0 0] [1 0 1] [0 0 1] [0 0 0]] [0 -1 0 0]].
] def
```

Assembling cubes by hand takes some care in order to get all the orientations correctly – even for this simplest of polyhedra. The faces here come in pairs that one might call back and front; the array on a back face is almost the same as that on the front but in reversed order and a shift in one coordinate. The face

```
[[[0 0 1] [1 0 1] [1 1 1] [0 1 1]] [0 0 1 -1]]
```

is the front face of the cube where $z = 1$, and

```
[[[0 1 0] [1 1 0] [1 0 0] [0 0 0]] [0 0 -1 0]]
```

is the back face corresponding to it, where $z = 0$. There are ways to use the symmetry of the cube to generate all its faces automatically, but this will not be explored in this book. In fact, one rarely constructs a polyhedron without some sort of organizational scheme in mind to generate it.

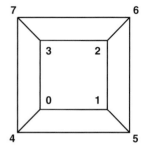

One thing that will make polyhedra more efficient is first to write out in a single array all of its vertices and then refer to this array in listing the faces. It is also a good idea to record the normal functions of the faces before drawing begins. For the cube, for example, it would be better to include the following code in your program rather than the stuff written above:

```
/V [
  [0 0 0] [1 0 0] [1 1 0] [0 1 0]
  [0 0 1] [1 0 1] [1 1 1] [0 1 1]
] def

/cube [
  [[V 4 get V 5 get V 6 get V 7 get] dup normal-function]
  [[V 3 get V 2 get V 1 get V 0 get] dup normal-function]
  [[V 0 get V 4 get V 7 get V 3 get] dup normal-function]
  [[V 5 get V 1 get V 2 get V 6 get] dup normal-function]
  [[V 0 get V 1 get V 5 get V 4 get] dup normal-function]
  [[V 7 get V 6 get V 2 get V 3 get] dup normal-function]
] def
```

It's easier to use this numbering in seeing how to get the face arrays oriented correctly, too.

One interesting exercise is now to write out the arrays representing the triangles of the faces of a **regular tetrahedron**, which is a 3D figure with four vertices and four faces, all of which are equilateral triangles. The hard part is to find its vertices, and there are several ways to do this. The first is to take a suitable selection of four vertices from a cube – a selection of cube vertices with the property that any two are separated by a diagonal on a face of the cube. I'll leave that construction as an exercise without further comment. The trouble with it is that you aren't quite sure where you are, so to speak; the tetrahedron you get has an orientation in space that also requires some calculation to work conveniently with. More convenient ultimately, but more difficult to design in the first place, is one with a fixed radius and orientation for which

■ the center is at $(0, 0, 0)$;

■ the top is placed at $(0, 0, 1)$;

■ this leaves us free to rotate the tetrahedron around the z-axis but now fix it by specifying one vertex of the bottom is at $(0, y, z)$ for some suitable values of $y > 0$ and z.

EXERCISE 14.1. *Figure out what the vertices of this tetrahedron have to be to make all faces equilateral. Compose PostScript code as was done above for the cube.*

EXERCISE 14.2. *Draw a regular tetrahedron by hand, including vertex numbers, and write a code fragment analogous to that above.*

EXERCISE 14.3. *A* **regular octahedron** has eight sides, all equilateral triangles. Find the vertices of some regular octahedron. Write PostScript code to draw it.

14.3   VISIBILITY FOR CONVEX POLYHEDRA

Usually when you draw objects in 3D they are solid – that is, a face blocks off the faces behind it. Now in some circumstances this can be quite difficult to deal with, but there is a large class of objects for which it is easy: polyhedra that are **convex** and **closed**. "Closed" means it has no holes. "Convex" means it bulges out at all points, or at least never bulges in. For example, a sphere is convex, but the surface of a doughnut is not because it has a hole in the middle. The difference, for drawing purposes, is this: if a closed object is not convex, then in some views you will have two faces pointing towards the eye, one at least partly obscured by the other. This means that you have to draw them in the correct order so that the one behind is covered over by the one in front. If an object has holes, the plates making it up should be considered to have two sides – literally an in-side and an out-side – and it will not be convex. We'll look at the problems of drawing nonconvex bodies in Section 14.7.

A convex surface will lie entirely on one side of the plane spanned by each of its faces. (This is actually the technical definition of convexity.) To draw a convex surface, therefore, we can just check visibility of each face separately without worrying about other faces. To check visibility for a single face, we have to determine whether the eye lies in the region of visibility of that face. Something subtle is involved in this. In constructing a face we calculate its normal function – $f = Ax + By + Cz + D$ for instance – and to check visibility for that original face we just have to evaluate that function on the eye. The eye is usefully expressed in homogeneous or 4D coordinates as $[x_e, y_e, z_e, w_e]$. To check visibility, we therefore just check the condition

$$[x_e, y_e, z_e, w_e] \bullet [A, B, C, D] = Ax_e + By_e + Cz_e + Dw_e \geq 0 .$$

But now in the course of making our program we have probably performed some transformations – for example, rotations – on the surface. This changes the points we draw and changes also the normal function of the face. In principle we can recover the coordinates of the transformed face and calculate also its new normal function. We can do this because one of the data structures in ps3d is the current 3D transform matrix T, which transforms from the coordinates we are currently working with to the original default 3D coordinates. But there is something more efficient to do. As far as visibility is concerned, rotating an object in one direction is equivalent to rotating the eye in the opposite direction. Similarly, moving an object away from you is equivalent to translating the eye away from the object in the opposite direction. In other words, if we want to check visibility of a face, we can *either* evaluate the visibility of the transformed face at the true eye *or* evaluate the original normal function at the eye-equivalent obtained by applying the inverses of the coordinate changes. For the first method, we would apply the current 3D transform matrix T to the original face. But the ps3d mechanism also holds the inverse of this matrix T^{-1}. So applying T^{-1} to the eye will tell us where the "virtual eye" is at any moment. The matrix T is item 0 in the array you get by calling cgfx3d, and T^{-1} is item 1 in it. Therefore, the code

```
/E get-eye cgfx3d 1 get transform3d def
```

defines E to be the virtual eye; the procedure get-virtual-eye in ps3d does exactly the same. The command sequence cim3d is also a shorter replacement defined in ps3d for cgfx3d 1 get. Here is a fragment of program that, given the fragment above defining a cube and the usual 2D stuff setting the scale, will draw only the visible faces of a cube:

```
[0 0 4 1] set-eye
[1 1 1 ] 60 rotate3d
-0.5 -0.5 0 translate3d

/E get-virtual-eye def

cube {
  aload pop
  /f exch def % f = normal function
  /p exch def % p = array of vertices on the face
  f E dot-product 0 ge {
    newpath
```

```
      % move to last point first
      p p length 1 sub get aload pop moveto3d
      p {
        aload pop
        lineto3d
      } forall
      stroke
    } if
  } forall
```

14.4 SHADING

Checking visibility of faces will contribute to an illusion of solidity – especially if the object is shown in a sequence of successive positions in a kind of animation. Another technique for creating an illusion of solidity is that of shading faces according to where they are located relative to an imaginary light source.

The light source will usually be a direction in 3D and therefore a 4-vector whose last coordinate is 0. A light source straight overhead would be [0 1 0 0], for example. Conventionally, a light source from overhead, slightly behind, and slightly to the left seems to be what the human eye is comfortable with, which would make the vector [-0.25 1 0.25 0]. It is best to normalize the light source so it has total length 1.

The shade of a face will then be a function of the angle between the light source and the vector perpendicular to the face. If it is $0°$, the face will be towards the light and bright. If it is $180°$, it will be away from the light and dark. The angle is a function of the dot product of the two vectors, and in fact there is no need to work with the angle itself – just the dot product.

With our normalizations of the normal function and the light source, the dot product will lie between -1 and 1. If -1 it will be dark and if 1 bright. The simplest way to assign a shade is to let d be the dot product and assign a color $(d + 1)/2$ in PostScript. So -1 becomes black and 1 white.

As with the eye, it is best to use a light-source-equivalent; thus, possible code for shading is this:

```
/light-source [-0.25 1 0.25 0] normalized def
/L light-source ctm3d 1 get transform3d def

cube {
  aload pop
  /f exch def    % f = normal function
```

```
/p exch def    % p = array of vertices on the face
f E dot-product 0 ge {
  newpath
% move to last point first
  p p length 1 sub get aload pop moveto3d
  p {
    aload pop
    lineto3d
  } forall
  gsave
  /s L f dot-product 1 add 2 div def
  s setgray
  fill
  grestore
  stroke
} if
} forall
```

This is not yet ideal. Relying on a straight linear translation from dot product to shade produces lighting that seems a bit harsh to the eye. For one thing, in the real world even the darkest places usually have a bit of reflected light from the environment, and so the darkest shade allowable shouldn't actually be black. And often you won't want the brightest to be 1, which will make a face invisible against a white background. It is best to allow more control over the translation – to allow a more general function to do the job. We'll think of this in the following way: we first calculate the linear translation $(d + 1)/2$ just as above, but then we fudge things a bit by translating the result, which lies between 0 and 1, to some other number between 0 and 1. We want to be able to specify a function from the interval $[0, 1]$ to itself. The best way to understand what is going on is to look at the graph of our fudge function. We can put it all in a unit box.

For example, here is the default translation, with no fudging (recall, we are looking at the translation after we have moved $[-1, 1]$ into $[0, 1]$):

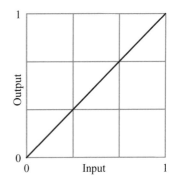

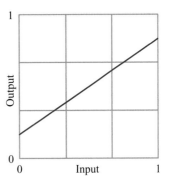

If we want to lighten the shadows and darken the bright spots, we want something like this:

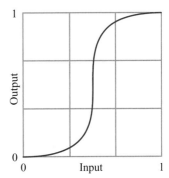

If we want to increase contrast, we want darks darker, lights lighter. If we want a large increase in contrast, we want an S-shaped curve like this:

The default method of fudging I have included in ps3d doesn't allow the full range of these options. It uses a procedure I call shade to translate from the dot product to a shade factor. There are two arguments to shade: a single number between -1 and 1 and an array of four numbers. The first and last numbers in the array are the minimum and maximum values of the shade factor, and the other two are more subtle parameters of the shade function, ones that determine control nodes $(1/3, s_1)$ and $(2/3, s_2)$ for the graph.

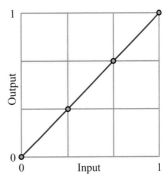

For the default shading, without fudging, the array is [0 1/3 2/3 1]:

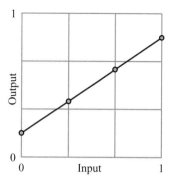

For an arbitrary straight line fudge function, choose the values in the array so that $(0, s_0)$, $(1/3, s_1), (2/3, s_2), (1, s_3)$ all lie in a straight line.

For nonlinear fudging, the control values s_1 and s_2 are chosen to force the fudge graph to lie close to the points $(1/3, s_1)$ and $(2/3, s_2)$ without passing through them.

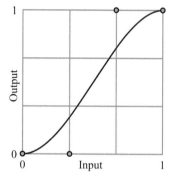

There is a fault with this scheme, since the amount of contrast it can achieve is limited:

To avoid unwelcome curiousities, the array numbers should satisfy $0 \leq s_0 \leq s_1 \leq s_2 \leq 1$, but anything in that range should be acceptable.

For more control over shading, say with higher contrast, you can apply a Bernstein polynomial of degree higher than 3. The disadvantage of doing this is that it requires more computational effort; the advantage is that you can make arbitrarily strong contrast.

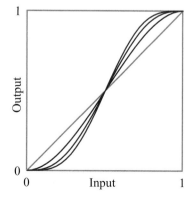

Here, for example, are some fudge graphs you can get by using a cubic Bézier curve (i.e., Bernstein polynomial of degree 3) and then Bernstein polynomials of degree 5 and 7.

14.5 SMOOTH SURFACES

In reality, most surfaces look essentially smooth and not at all like polyhedra. Nonetheless, computers can only approximate them – usually as polyhedra together with some extra data. For a few surfaces like spheres there are special ways they can be approximated in this manner, but for most surfaces the first step in drawing them is to **parametrize** them – to find maps from 2D into 3D that describe them.

Let's start with spheres. A sphere is determined completely by its center and its radius. In drawing, the center might as well be taken to be the origin since `translate3d` can just move a sphere around. Different radii can be dealt with by `scale3d`, and so I'll assume the radius to be $R = 1$.

A point on the surface of a sphere has **longitude** θ and **latitude** φ as its coordinates. Longitude measures the angular distance between a point and a fixed meridian line, whereas latitude measures how far a point is from an equator. If a point on a sphere of radius R has longitude θ and latitude φ, then its (x, y, z) coordinates are

$$S(\theta, \varphi) = (R \cos \theta \cos \varphi, \, R \sin \theta \cos \varphi, \, R \sin \varphi) \, .$$

The parametrization must have the right orientation; that is, these variables must look more or less like x and y everywhere on the surface as it does here. The parametrization procedure has two arguments θ and φ and returns the 3D point $S(\theta, \varphi)$.

```
% longitude latitude -> point on unit sphere
/P { 1 dict begin
  /l exch def
  /L exch def
   % L=longitude l=latitude
  [L cos l cos mul
   L sin l cos mul
   l sin ]
end } def
```

We now want to use this parametrization to approximate the sphere by an assembly of flat plates. That's easy – a plate will be what you get when you map a coordinate rectangle into 3D. As θ ranges from 0 to 360 and φ from -90 to 90 (using degrees instead of radians, as is normal in PostScript), we cover the whole sphere. First we assemble the vertices in a double array. We use a procedure with a single argument N that returns an N by $2N$ array of grid points on the sphere except that the poles are singletons. Each internal array is an array of points laid out along a parallel of latitude.

```
% N -> latitudes in N+1 rows
/sphere-vertex { 1 dict begin
/N exch def

/dA 180 N div def
/dB 360 N div 2 div def

 % A = latitude
[
  [
    [0 0 -1]
  ]
/A -90 dA add def
N 1 sub {
  [
    /B 0 def
     % B = longitude
    2 N mul 1 add {
      B A P
      /B B dB add def
    } repeat
  ]
  /A A dA add def
} repeat
[
  [0 0 1]
]
]
end } def
```

Finally, we assemble the vertices into faces.

```
% N -> array of faces
/sphere { 1 dict begin
/N exch def

/S N sphere-vertex def
 % S now is an array of vertices, arranged in latitudes
```

```
[
    % the triangles at the south pole
  0 1 2 N mul 1 sub { /j exch def
    [
      [ S 0 get 0 get
        S 1 get j 1 add get
      S 1 get j get
      ] dup normal-function
    ]
  } for
   % the rectangular regions in the middle
  1 1 N 2 sub { /i exch def
    0 1 2 N mul 1 sub { /j exch def
      [
      [
      S i get j get
      S i get j 1 add get
          S i 1 add get j 1 add get
          S i 1 add get j get
        ] dup normal-function
      ]
    } for
  } for
    % the triangles at the north pole
  0 1 2 N mul 1 sub { /j exch def
    [
      [ S N 1 sub get j get
        S N 1 sub get j 1 add get
        S N get 0 get
      ] dup normal-function
    ]
  } for
]
end } def
```

To draw the regions on the sphere that we have constructed, we can just plug in the array of faces of a sphere where we dealt with the faces of a cube in earlier code.

```
/S 18 sphere def
/E get-eye cim3d transform3d def
/L light-source cim3d transform3d def
```

```
S {
  aload pop
  /f exch def    % f = normal function
  /p exch def    % p = array of vertices on the face
  f E dot-product 0 ge {
    newpath
    % move to last point first
    p p length 1 sub get aload pop moveto3d
    p {
      aload pop
      lineto3d
    } forall
    gsave
    L f dot-product [ 0.2 0.2 0.9 0.9 ] shade
    setgray
    fill
    grestore
    stroke
  } if
} forall
```

Here is what we see with $N = 16$ with the rectangle boundaries stroked in one version but not in the other.

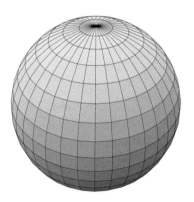

In the following figure $N = 32$. I would have liked to have included a picture with $N = 64$, but that would have meant drawing – well, making an attempt to draw – roughly $8,192$ rectangles in 3D. I am afraid that my printer freezes up at the prospect. We'll see in the next section that this is unnecessary anyway.

14.6 SMOOTHER SURFACES

Version 3 of PostScript introduced high-quality shading. In earlier versions, if you wanted to show a gradient of color across a region you had to fill in all the separate colors in small regions or maybe draw zillions of lines of different colors. This was almost always painful, slow, or both. In the new scheme, you set colors at certain points and PostScript fills in between these points by interpolating the specified colors. There are several different methods for doing this – seven in all, to be exact – but we will look here at one called **free-form Gouraud shading** that nicely balances simplicity against quality. It fits in quite well with surface parametrization. There is one extra requirement now, however: the parametrization must specify not only location but also the normal function at each point of the surface. We'll see why in a moment. At any rate, this is easy enough to obtain from the parametrization by calculus. If the parametrization is

$$(s, t) \longmapsto (x(s, t), y(s, t), z(s, t)),$$

then the vectors

$$[\partial x/\partial s, \partial y/\partial s, \partial z/\partial s], \ [\partial x/\partial t, \partial y/\partial t, \partial z/\partial t]$$

span the tangent plane to the surface and their cross product will be perpendicular to it and facing out – as long as the orientation of the parametrization is correct – that is, so that (s, t) have the same orientation as (x, y) on the surface. Normalize this cross product to get $[A, B, C]$. For Gouraud shading, the parametrization should return this normal vector as well as the location. For the sphere, the following procedure will do because the normal vector is the same as the location vector:

```
% longitude latitude -> [ point on unit sphere, normal vector ]
/P { 1 dict begin
```

```
/l exch def
/L exch def
 % L=longitude l=latitude
/x L cos l cos mul def
/y L sin l cos mul def
/z l sin def
[[x y z] [x y z 0]]
end } def
```

The way PostScript does shading of any kind is through a special data structure called a **shading dictionary**. All you have to know is that the code that produces the shading looks like this:

```
<<
  /ShadingType 4
  /ColorSpace [ /DeviceGray ]
  /DataSource [ 0 x y g ... ]
>>
shfill
```

or this:

```
<<
/ShadingType 4
/ColorSpace [ /DeviceRGB ]
/DataSource [ 0 x y r g b ... ]
>>
shfill
```

Here shfill is a command, a "shading fill," with the dictionary << ... >> as argument. A dictionary is a list of **key** and **value** entries that associates a value with each key word listed. In the dictionary /ShadingType 4 specifies what kind of a shading dictionary (free-form Gouraud) this is. Another kind that you might want to look into is Type 6, **Coons patch meshes**, which allows you to introduce curvature into your shaded fragments. These are all discussed in the section on patterns in the PostScript Reference Manual. The key ColorSpace specifies the type of colors to be used. The value /DeviceGray says that colors are specified by a single number between 0 and 1, which specifies a shade of gray. Another possibility would be /DeviceRGB, indicating three numbers *RGB*. The /DataSource lists 2D points and colors to be interpolated between them. Here it is an array of 3×4 numbers since there are three vertices in a triangular plate. Each of the four numbers is associated to a single vertex. The first of the four numbers is a "magic number" that for us will always be 0. The next two are coordinates in the current-user 2D coordinate system of the image of the vertex, and g the shade of gray associated to that vertex. If the color space is RGB, then this single number becomes three color components. The point (x, y) is calculated from our 3D points in a way I'll explain in a moment.

These shading routines are not unique to 3D drawing. The following code fragment draws a single triangle with colors black, red, and white at the vertices.

```
/A [ 0 3 sqrt 2 div ] def
/B [ -0.5 0 ] def
/C [ 0.5 0 ] def

/ds [
  0 A aload pop 1 0 0
  0 B aload pop 1 1 1
  0 C aload pop 0 0 0
] def

newpath
<<
  /ShadingType 4
  /ColorSpace [ /DeviceRGB ]
  /DataSource ds
>>

shfill
```

For a 3D object, the color at any point will be determined by the light source and the normal vector at that point. Thus, the data structures needed for Gouraud shading are slightly different from the flat-plate scheme we used earlier. For one thing, *we now must use triangular plates instead of rectangular ones*. To use Gouraud shading to draw any 3D shape, the first step is to build it as a family of triangles, each with normal vectors at the vertices plus a normal function for the triangle itself. The normal vector for the triangle is used to test visibility, and the ones at the vertices are used to interpolate colors. We think of the surface to be drawn as an array of such colored triangles.

To be precise, a **vertex** P here is an array [[x y z] [A B C 0]], where x, y, z are the 3D coordinates of P, and $[A, B, C]$ is the unit normal vector at P. This is the sort of structure to be returned from the parametrization procedure. A **triangle** is an array [P Q R [A B C 0]], where P, Q, R are vertices in this sense. A **surface** is an array of triangles. Thus,

```
/T [
  P Q R
  [ P 0 get Q 0 get R 0 get ] normal-function
] def
```

defines a triangle T if *P*, *Q*, *R* are vertices.

Usually these data will be constructed from a parametrization, but not always. We'll discuss how to do that efficiently in a moment.

Given an array `surface`, an array of triangles in this sense, here is how it is drawn:

```
surface {
    % [ P Q R normal ] now on stack
  aload pop
  /f exch def    % f = normal function
  /R exch def
  /Q exch def
  /P exch def
   % P, Q, R  = vertex = [ pt + normal ]
  f E dot-product 0 ge {
    newpath
      % define grey tones for shading
    /sP L P 1 get dot-product 1 add 2 div def
    /sQ L Q 1 get dot-product 1 add 2 div def
    /sR L R 1 get dot-product 1 add 2 div def
    /ds [
    0 [ P 0 get aload pop 1 ] CTM transform3d render sP
    0 [ Q 0 get aload pop 1 ] CTM transform3d render sQ
    0 [ R 0 get aload pop 1 ] CTM transform3d render sR
  ] def

newpath
<< /ShadingType 4
   /ColorSpace [ /DeviceGray ]
   /DataSource ds >>
     shfill
  } if
} forall
```

The line

```
0 P 0 get transformto2d sP
```

has the "magic number" 0; the 3D point P rendered to 2D; and then the shade at P. Here is what the figure looks like if the surface is a sphere:

Smooth, eh? Shading in this figure has been done with a Bernstein polynomial of degree 5.

Most surfaces will be built from a parametrization by a rectangular array. Here is a sketch of the way things go – at least when we are drawing a surface covered by a rectangle via parametrization. Before we do any drawing at all, we do some preparation:

(1) We write the parametrization function, which returns an array of two arrays, location plus normal function.

(2) We build the $(M + 1) \times (N + 1)$ array of the object's 3D points plus normal function at those points by using the parametrization function. Here M and N are positive integers, which we choose large enough to give an illusion of smoothness. Experimentation will probably be necessary to get them right. In effect we have now built a grid of size $M \times N$ covering the surface.

Then we build all the triangular faces we want to draw, using the grid as indicated here:

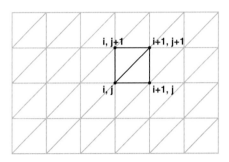

Here's code that does this:

```
/sphere [
0 1 2 N mul 1 sub {
  /i exch def
   0 1 N 1 sub { /j exch def
   /P S i get j get def
   /Q S i 1 add get j get def
   /R S i 1 add get j 1 add get def
   /n [ P 0 get Q 0 get R 0 get ] normal-function def
   n length 0 ne {
        [ P Q R n ]
} if
/P S i get j get def
/Q S i 1 add get j 1 add get def
/R S i get j 1 add get def
/n [ P 0 get Q 0 get R 0 get ] normal-function def
n length 0 ne {
   [ P Q R n ]
  } if
 } for
} for
] def
```

14.7 ABANDONING CONVEXITY

Few scenes other than the very simplest, even in theoretical mathematical figures, consist of a single convex body. A simple visibility test by means of normal functions will no longer work.

The basic problem can be demonstrated with a simple example. Suppose that you want to draw not just one but two cubes. In certain situations, one of them will hide part or even all of the other from view.

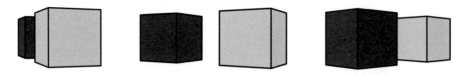

The basic idea is to draw the one that is farthest away first and then the nearest one because PostScript paints over what it draws. Indeed, this strategy is called the **painter's algorithm**. Furthermore, if you are doing an animation, in effect moving your eye around the pair of cubes, which one is farthest away and which one is nearest will change, and so you must decide dynamically as the scene changes which is to be drawn first. In addition, if you have a large number of objects to draw, you will have to make these dynamical choices efficiently. These all seem like impossible demands, right?

Drawing complicated scenes is much more involved in 3D than in 2D – and more interesting since some real ideas are required. Much high-end 3D drawing, for example in video games or movies, relies on a pixel-by-pixel treatment. The pixels in hardware designed for this purpose incorporate a **depth** coordinate – that is depth with respect to the plane of the screen – and pixels are colored in the order of their depth, which means that close pixels are painted after far ones. This hardware option is unavailable to PostScript, which is essentially device independent. The PostScript program itself must therefore be responsible for keeping track of depth. The standard method for doing this is to use a **binary space partition**.

In drawing the two cubes, for example, the trick is to place a plane between the two cubes and use that to keep track of how the eye is related to the cubes. The cube to draw first is the one on the side of this plane opposite the eye.

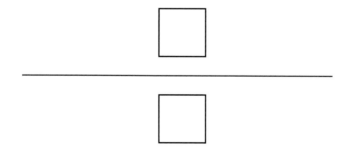

In other words, we divide space into two parts by a plane. Each side of this plane contains one of the cubes. The question of nearest and farthest is determined by ascertaining which side of this plane the eye (or, in practice, what I have called the **virtual eye**) is on.

This division of space into two components by means of this plane is about the simplest example of a binary (two-fold) partition of space. If more than two convex objects are to be drawn, the idea is to partition all of space by a plane separating one group of objects from the other and then to partition each of the half-spaces if necessary by its intersection with a plane, and so on, until each of the pieces that space has been chopped up into contains exactly one of the objects to be drawn.

This partition is stored in a way that makes it easy and rapid to decide where the eye lies in relation to these partitioning planes. The important thing is that the construction of the partition itself, which is a relatively time consuming, can be accomplished before any drawing at all. The amount of work involved in this is roughly proportional to n^2 if n objects are to be drawn. The drawing itself turns out to be proportional to n, which is generally much smaller.

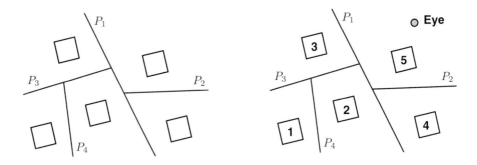

The basic idea, illustrated on the left, is first to partition all of space into two pieces by a single plane (here P_1) and then partition each of the remaining components by its intersection with a plane (P_2 and P_3), and so on. When using this partition to draw, in this example the program first decides what side of P_1 the eye lies on and then recursively looks at each side in turn. The right-hand image shows the order of drawing with the eye at upper right. The structure necessary to do this drawing is a binary tree whose nodes are the separating planes and whose leaves are the objects to be drawn.

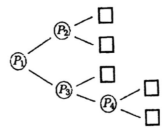

This strategy of partitioning by half-spaces might work easily, or it might not. Sometimes some extra work is involved. As the figure on the left below illustrates, the given figures might have to be chopped up. This is acceptable because what we are drawing are surface fragments, and a plane cuts a surface fragment into smaller fragments. But we definitely have to allow for this chopping.

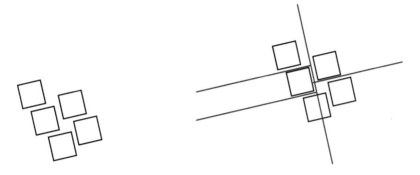

There are special ways to draw special surfaces, but a binary space partition can be used to draw any collection of surface fragments from any perspective. To be precise, a binary space partition is a tree, as I have mentioned, with two kinds of nodes, branches and leaves, and it is defined in an essentially recursive manner. A leaf is simply a surface fragment. A branch is in essence a separating plane, and in practice this plane is chosen to be the support of one of the surface fragments. So it becomes a surface fragment together with the two binary space partitions of the half spaces it determines, which conventionally are called left and right. If $f \geq 0$ is the defining equation of the separating plane, then all the leaves of the left partition lie entirely in the region $f \leq 0$ and all those on the right lie in $f \geq 0$. More completely, any leaf lies in the intersection of all of the half spaces associated to the branches containing it, which is of course a convex set. The leaves of a binary space partition are drawn recursively. Each node v of the partition is examined; if the eye lies, say, in the region $f_v \geq 0$, then all the leaves in the left partition are drawn, followed by the fragment associated to the node itself, followed in turn by the right partition. In constructing the partition associated to a collection of fragments, a fragment is chosen at random from the collection and all the other fragments split into the halves determined by its support. These are collected into left and right collections, and the partitions associated to each of these is then constructed. This is allowable because the chosen surface fragment itself is excluded from both left and right and the complexity of each partition construction is decreased at each step.

The results can be quite pleasant, although PostScript requires a long time to draw complicated figures.

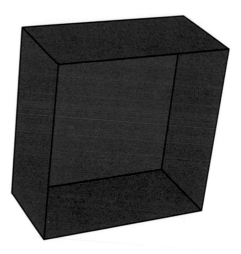

Certain configurations illustrate that, in some configurations, there is absolutely no way to guarantee back-to-front drawing without splitting up fragments since there is no way to order the original fragments.

Another application of the splitting algorithm used in binary space partitioning is to chop away things in 3D that lie behind the eye, thus eliminating the weird effects mentioned in Section 13.1.

14.8 SUMMARY

Drawing surfaces is quite complicated, and it is probably a good idea if I summarize the process here, at least in the most common cases. I'll work with the most complicated case in which we are interested using shfill. It should be easy enough to adapt what I say to the easier case that doesn't use it.

■ I'll assume the surface can be parametrized by a map $f(s, t)$ from 2D to 3D. The first step is to write the **parametrization function** f that has an array [s t] of length two as argument and returns the four item array [x y z n], where

n is the unit normal vector to the surface at (x, y, z), the point parametrized by (s, t). I'll call such an array a **vertex**.

- Build an array of vertices over the range you want to look at. Usually this will correspond to a rectangle in parameter space.

- Assemble the plates of the surface into an array. These can be triangles, rectangles, and so on. It is best – that is, most efficient – to build the largest flat plates possible. For example, on a sphere you build the rectangles laid out by longitude and latitude. Be careful about singular points of the parametrization, such as the poles of the parametrization of the sphere by latitude and longitude, where the rectangles collapse to triangles.

- Decide on a color scheme – usually just a single color.

- Pick a light source and a shading scheme.

- If you're dealing with a single convex object, there is nothing more to be done except to draw it and shade it, testing visibility with the normal function for plates.

- Otherwise, make up a BSP structure. Then draw. In my code, the procedure that builds the BSP has as one of its arguments an interpolation routine that constructs the normal vector at a point on a segment between two other vertices. This is needed because constructing the BSP tree occasionally requires that the plates it starts with get split into smaller plates.

14.9 CODE

PostScript routines for building binary space partitions can be found in the file `bsp.inc`, and sample usage in `triad.ps`, `box.ps`, and `doughnut.ps`. The last file illustrates a complete construction of a smooth surface. Well, maybe not that smooth – note the straight line segments on the boundary of the doughnut. This can be fixed by only the more sophisticated shading using Coons patches.

The parametrization of the doughnut (or **torus**, which my Latin dictionary translates as "cushion") is

$$(s, t) \longmapsto ((R + r \cos t) \cos s, (R + r \cos t) \sin s, r \sin t),$$

where R and r are the two radii involved. The easiest way to see this is to start with the circle in the (x, z) plane of radius r at center $(R, 0)$ and rotate it around the z-axis in 3D.

In implementing binary space partitioning for drawing fragments, some extra care has to be taken to avoid nasty floating point problems. The main difficulty is in splitting, for if two fragments are almost parallel and one is split by the other, the result may vary wildly in response to floating point errors in the calculation. In practice, in drawing 3D figures many fragments will have exactly the same normal function, which can cause extremely bad effects without precautions. One good thing to do is to assign all these fragments *exactly* the same normal functions – that is, exactly the same array. Another is not to split any fragment by a plane approximately parallel to it, although this is a costly and somewhat arbitrary test.

REFERENCES

1. M. de Berg, M. van Kreveld, M. Overmars, and O. Schwarzkopf, **Computational Geometry – Algorithms and Applications**, Springer-Verlag, 1991. Chapter 12 is all about binary space partitions.

CHAPTER 15

Triangulation: basic graphics algorithms

We now know how to draw parametrized surfaces in a reasonably realistic manner. We do not yet, however, know exactly how to draw even simple regions on surfaces, such as spherical triangles, if we want to take into account the usual phenomena of shading and visibility.

The techniques explained for managing the whole surface suggest that the natural way to proceed is to chop up the region involved into small triangles and then apply the surface parametrization to those triangles. The problem is thus reduced to that of decomposing an arbitrary 2D polygonal region into small triangles, and that will be the principal topic in this chapter. This is, as one might suspect, a very basic problem in computer graphics, and there are several valid approaches. The one taken here can be found in Chapter 3 of the book by de Berg et al. that I have already mentioned as the source of the algorithm for finding convex hulls in 2D as well as for dealing with binary space partitions.

What is to be explained is the most complicated single topic in this whole book. It involves among other things some very basic computational structures we have so far managed to avoid such as stacks and binary search trees. Implementing these in PostScript is . . . hmmmm . . . educational, Intriguing, Bracing, And still, overall, valuable. Using these and other basic data structures is common in sophisticated graphics programs because much information has to be stored accessibly to avoid having to go backwards in a calculation as we go along, and this can be tricky.

Practical triangulation involves three steps: decomposition into vertically monotone regions, then into triangles, and then into small triangles. None of these is uninteresting.

15.1   THE MONOTONE DECOMPOSITION

I'll assume that we are working with a simple planar region, one whose boundary is a simple, closed polygon. I'll also assume for purely technical convenience that all edges are nondegenerate, which means that no two successive vertices are the same. Orient the boundary according to the right-hand rule so that the interior of the region lies on its left bank (using here the geographical terminology applied to rivers).

A region is described as being **monotone** (or more precisely **vertically monotone**) if its boundary descends on its left side and ascends on its right. In the exceptional case that its boundary possesses horizontal fragments, a horizontal segment moving left to right is considered to be **descending**, and one moving right to left to be **ascending**. A region fails to be monotone if traversing the boundary on its left or right side involves some reversals in vertical direction. Roughly speaking, a vertically monotone region is one with the property that any horizontal slice, maybe rotated counterclockwise slightly, always meets its boundary in two or fewer points. The figure at right, for example, is clearly not monotone because some horizontal slices intersect it in more than two points.

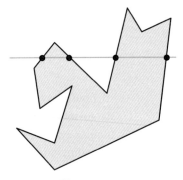

Nonmonotone

As we will see in Section 15.4, a monotone region is relatively simple to triangulate. The principal step in the whole process, and the one that takes the most time in a technical sense, is chopping the region into monotone pieces. How to do this will turn out to be a rather intricate business.

The basic idea in the monotone decomposition is to perform a top-to-bottom sweep of the region by a horizontal line. As the **sweep line** is moved down, something happens whenever it meets a vertex of the boundary, adding a diagonal occasionally from the vertex encountered to some other vertex encountered before and updating the structures necessary to doing this. The updating involves keeping track of the descending polygon segments intersecting the sweep line as well as of the vertices last encountered just to the right of them. Adding diagonals means splitting regions into halves. Here is how the process goes for the region above:

  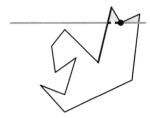

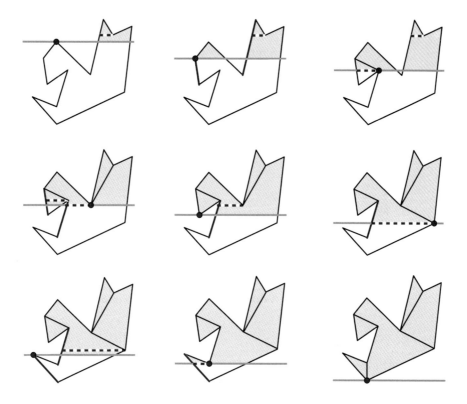

The figures exhibit some extra data that I'll explain in this Section.

As the sweep proceeds, it jumps from one vertex to the next lower one. What it does at that point depends on the type of vertex it encounters. There are six possibilities according to the relative position of the entering and leaving vectors u and v. Call a vector $[x, y]$ **ascending** if $y > 0$ or $y = 0$ and $x < 0$, and call it **descending** otherwise. Let $u^\perp$ be u rotated counterclockwise by $90°$.

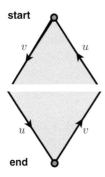

A **start** vertex. Here u is ascending, v descending, and $v \cdot u^\perp > 0$.

An **end** vertex. Here u is descending, v ascending, and $v \cdot u^\perp > 0$.

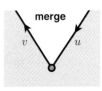

A **merge** vertex. Here u is descending, v ascending, and $v \bullet u^{\perp} < 0$.

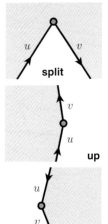

A **split** vertex. Here u is ascending, v descending, and $v \bullet u^{\perp} < 0$.

An **up** vertex. Here u and v are both ascending.

A **down** vertex. Here u and v are both descending.

The classification might be clearer from these pictures:

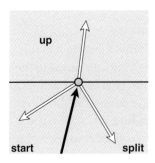 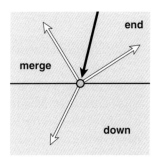

A region is monotone precisely when it has no merge or split vertices, and so the point of the process is to draw diagonal lines to eliminate them by separating the polygon into smaller pieces. The difficulty is to know when and how to do this. Two basic principles are involved in carrying out the sweep:

(1) Whenever we encounter a split vertex we must draw a diagonal line upwards from it;

(2) whenever we encounter a merge vertex we must mark it and be sure to draw a diagonal line upwards to it eventually. Thus, every diagonal drawn will have either a merge or a split vertex at one end, possibly both. Knowing when to

add one will require keeping track of several different things as we go, and the extra data are indicated in the figures above.

First of all, we must keep a list $\mathcal{L}$ of all the descending boundary segments encountered by the current sweep line. The important thing is to be able to determine for any vertex what boundary segment lies immediately to its left, and this can change quite drastically as we move down. The list $\mathcal{L}$ will be have to be ordered and searchable as well as dynamically maintained.

Second, diagonals will always be drawn back from a vertex we have just met to one we encountered earlier. Geometrically, it is easy to describe what the earlier one is. If e is any descending segment of the polygon, its **helper** at any given moment is the vertex u that can be connected to e across the interior of our region lowest among those above the sweep line with this property.

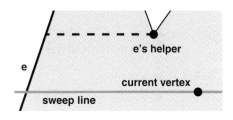

When a helper is a merge vertex, it is a candidate for the top of a diagonal to be drawn from below; when it is a split vertex, a diagonal must be drawn upwards from it. Each descending boundary segment in the sweep intersection list is associated to its helper, and this assignment may change in the course of the sweep. The crucial property of a helper is that the region bounded by the sweep line, the descending segment e (just to the left of the current vertex), the line to the helper, and the diagonal from the current vertex to the helper is always a region free of other vertices and edges.

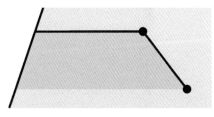

What about the ordered list of intersections? It will be a **search list** with dynamic data. At any moment it is to store all the descending boundary segments intersecting the sweep line ordered left to right according to the intersection. There are three operations we must be able to perform on it:

- remove a given edge;
- insert an edge in the correct place;
- find the edge in the list immediately to the left of a given vertex. It is the last operation that is the crucial one, but maintaining this property correctly requires the previous two.

In addition to this dynamic list, we must make up a list – a static queue – of all vertices of the polygon right at the beginning of the sweep, ordered according to height, top to bottom. This requires an initial sort and then keeping track of vertices already dealt with. In principle, it is this sort that sets one lower bound to the complexity of the whole process, for if there are n vertices it will generally require $O(n \log n)$ time.

15.2 THE ALGORITHM

Start with the array p of 2D points representing the vertices p_i of our simple closed polygon. We'll assume there to be at least three. This array will be the argument to the triangulation algorithm. The output will be an array of polygons of the same type making up the monotone decomposition.

We build the vertex list with each (enhanced) vertex now an array of

- a point p_i
- a predecessor edge
- a successor edge
- a type
- an index.

The index is the index of the vertex in the array of vertices making up the original polygon. It is needed for technical reasons when making up the polygons that go into the output. Deciding the type of a vertex will require looking at predecessor and successor edges. At the same time, we build the edges, where an edge is an array of

- a starting vertex
- an end vertex
- the function $Ax + By + C$, which vanishes on the edge and is positive on the inner side
- the edge's helper
- its node it corresponds to in the list $\mathcal{L}$, if any.

The function $Ax + By + C$ is used to compute the intersection of the sweep line with the edge. When an edge is first created from a pair of vertices, its helper is null as is its node pointer.

We sort the vertex list, top to bottom, left to right in case of matches. Put this sorted list into a queue. Initialize the dynamic search tree $\mathcal{L}$ with (as I have already said) methods for insertion of an edge according to its position, removal of an edge by designation of its node in the tree, and location of an edge just to the left of a given vertex.

Although this queue isn't empty, keep popping a vertex v off the queue. According to what type this vertex is,

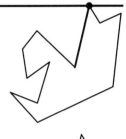

Start. Put the successor edge of v in $\mathcal{L}$. Set its helper equal to v.

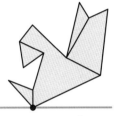

End. Let e be the predecessor edge. If its helper u is a merge, make a diagonal from v to u. Remove e from $\mathcal{L}$.

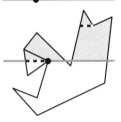

Split. Let e be the edge immediately left of v. Make a diagonal from v to its helper and set the new helper of e to be v. Add the successor edge of v to $\mathcal{L}$.

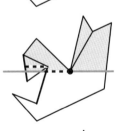

Merge. Let e be the predecessor of v. If u is its helper and it is a merge, make a diagonal from v to u. Remove e from $\mathcal{L}$. Let f be the edge in $\mathcal{L}$ directly left of v. If its helper w is a merge, make a diagonal from it to v. Set the new helper of f equal to v.

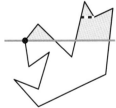

Down. The region lies to the right of v. Let e be the predecessor of v. If its helper u is a merge, make a diagonal from it to v. Remove e from $\mathcal{L}$. Insert the successor edge in $\mathcal{L}$ and set its helper equal to v.

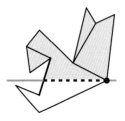

Up. The region is to its left. Let e be directly left in $\mathcal{L}$. If its helper is a merge, make a diagonal from it to v. Set the helper of e equal to v.

Making a diagonal means splitting a polygonal region into two pieces. This is done as these figures indicate:

The partition therefore adds two new vertices to the collection and reassigns predecessor and successor edges. Some care has to be taken to be sure when calling this procedure that assignment of helpers is compatible with this split.

It is an interesting exercise to check that this all works well – diagonals never intersect edges or other diagonals, and every split or merge vertex is fixed up sooner or later.

15.3   THE INTERSECTION LIST

As part of the process of finding a monotone decomposition, we need a list $\mathcal{L}$ of descending edges intersected by the current sweep line ordered by the x-value of the intersection. This list changes as the sweep line moves down. Three procedures are required of it. We must be able to

■ insert an edge into the list;

■ remove an edge;

■ find the edge just to the left of any given value of x

How can we implement such a list in PostScript? The simplest thing to do is maintain an array of all the current edges ordered left to right. But then all three operations would be quite inefficient. If n is the length of the current list, then locating an edge would on the average require $n/2$ comparisons, and inserting an edge would mean on the average shifting $n/2$ edges one place forward. Maintaining a **linked**

list with nodes holding both the data to be stored as well as a link to the next node would improve insertion and removal somewhat, but location would still be poor. The solution I use is a **binary search tree**. This is a tree whose nodes contain data as well as links to two descendants and a parent node. A good reference for this, or at least for a search tree closely related to the one we'll use, is Chapter 14 of the book by Sedgewick listed at the end of this chapter.

Often, the nodes of a binary search tree store data associated to a numerical key. Each node of the tree has left and right descendants, left and right, which may be null. Each of a node's left descendants has a key value less than its own, and each of its right descendants has a key value greater than or equal to its own.

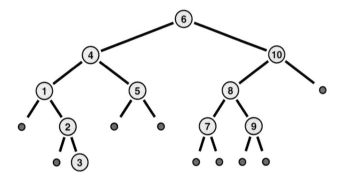

The point of a binary search tree is that we don't have to search through the whole list to deal with its entries but instead just have to go down through the branches of the tree. The first is an $O(n)$ process, the second on the average $O(\log n)$. In our case, each node of the tree will store a descending edge, and the nodes are sorted in the tree according to the x-value of an edge's intersection with the current sweep line. Explicitly, if the equation of the edge is $Ax + By + C = 0$ with $[A, B]$ pointing inwards, and y is the height of the sweep line, then

$$x = -\frac{By + C}{A}$$

if $A \neq 0$; otherwise, (the edge is horizontal) the x-coordinate of the top vertex of the edge. Insertion of data in such a tree is simple, as is location of a predecessor, but removal is tricky. Refer to Sedgewick's book for details of how it works.

There is something a little unusual about this search tree. The sweep line changes continually, and the key value of a node changes with it. This turns out not to matter because, although the key value of a node changes, the ordering of the current nodes does not change except with respect to a new insertion or removal. In the implementation we use, the search tree is assigned a procedure **key** that assigns to each node its x-value based on the current value of y.

15.4 TRIANGULATION

Once we have decomposed the original region into monotone subregions, the next step is to triangulate them. This is reasonably straightforward. Again I follow the book by de Berg et al.

The basic principles of the algorithm can be formulated succinctly:

- *Go from top to bottom.*
- *Make a triangle whenever possible.*
- *Cut away dead wood.*

Let's see how these work out in practice at the beginning of a triangulation. We start with a single point, the top of the polygon. Since the region is monotone, the next point has to lie along an edge from the first. So the starting point always looks essentially like this (up to left–right interchange):

When we place a third point, however, there are three rather distinct possibilities:

Following the principles laid out above, we obtain the following result:

The interesting thing is that in the first two cases we are essentially back at the starting point. In each of those, a triangle has been cut off, one point has become dead, and we are again looking at two active points that form vertices of a shrunken region. Whereas in the third case we are in a new situation with three points still active, their fate as yet undetermined. It is easy to see by an inductive argument that this is essentially what always happens in moving down one level – either we find ourselves in a simpler situation, or we find ourselves facing a fan configuration like that here on the left:

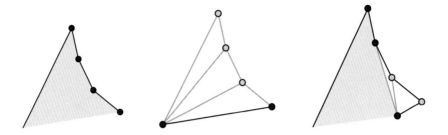

This is basically the same problem we saw previously. There will be one of two possibilities: the next point to be scanned lies either on the left, and we connect it to all the active points on the right (except the top, which is already connected), or on the right, and we connect it to all the active points that are visible to it (leaving active the top point connected as well as all above it). In any event, we update the list of active points and find ourselves facing a similar situation for the next step.

Here is a more detailed account of how things go. We start off by ordering the vertices, top to bottom, classifying them according to whether they lie on the left or right side of our polygon. This is possible because of monotonicity. Also because of monotonicity, this ordering can be done by starting at the top point and merging left and right sides by going backwards and forwards from the top. Then we set up a **stack** of active points. For the uninitiated, I recall that a stack is one of the most basic of all structures in computation, machine or human; items on a stack represent work postponed. Items on a stack are placed there and removed according to **LIFO** (**L**ast **I**n, **F**irst **O**ut) protocol. Putting something onto the top of the stack is called **pushing it**, and removing it from the top is called **popping** it. When you put a hammer down to go find a nail, you are pushing the use of the hammer onto your personal stack, and when you come back with the nail you are popping it. The procedures I use to implement a stack in PostScript are discussed at the end of this chapter.

We now put the two top vertices on the stack and start examining the remaining vertices, top to bottom. A vertex stored on the stack holds its 2D coordinates as well as the designation of what side it lies on. Suppose we are looking at the vertex v_i. Let u be at the top of the stack. What we do next depends on the exact situation:

(1) The vertex u is on the same side of the polygon as v_i. Pop u. Pop other vertices from the stack as long as they are visible within the region from v_i. Make triangles from these and their predecessor. Push the last one popped and v_i onto the stack.

(2) The vertex is on the opposite side. Pop all vertices from the stack. Make up diagonals for all but the last one. Push u and v_i onto the stack.

This works because there are always at least two vertices on the stack. Here is how the whole process goes for one region:

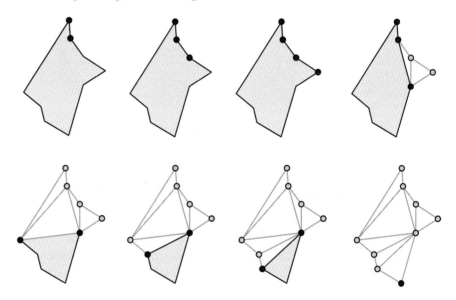

15.5 SMALL TRIANGLES

The triangles produced so far will vary considerably in shape and size. The last step is to produce from them a collection of uniform small size. How to do so elegantly depends on the shape of the triangle.

The mathematical problem can be formulated quite precisely: *How do we partition a given triangle into a minimum number of triangles of no more than some specified diameter δ?* This looks like a very difficult problem. If the diameter concerned is very small and hence a huge number of small triangles would be necessary, the solution likely involves paving a large part of the triangle with equilateral triangles. In practice, however, the diameter will not be all that small and it doesn't seem to me worthwhile to try for the best solution. So we look instead at the problem of partitioning a triangle into a reasonable number of smaller triangles of diameter no more than δ.

We can get some idea of what we are facing by first sorting the lengths of the sides, and so we may assume $a \leq b \leq c$. If we orient the triangle so its longest side lies along the interval $0 < x \leq c$ and reflect if necessary, we may assume the third vertex to lie in the region shown at the right:

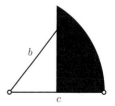

There are several extreme possibilities. Of course, if $c \leq \delta$ we do nothing. Otherwise, the triangle is going to partition into successively smaller ones in a series of steps until the diameter is at most δ, and the principal problem is to decide what one step amounts to. Somewhat arbitrarily, I choose the following procedure, which guarantees that at each step I reduce the diameter of the triangles being considered by a factor of $1/\sqrt{2}$:

■ *If $b/c \leq 1/\sqrt{2}$, we make a single cut:*

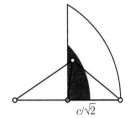

■ *Otherwise, if $a/c \leq 1/2$, we make two cuts:*

■ *In all other cases, we make three:*

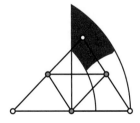

In the final algorithm, we use a stack to hold triangles that remain to be subdivided. Although the stack isn't empty, we pop triangles off the stack and subdivide them. Any triangles that are small enough are output, whereas those that are not are put back on the stack. It is most efficient if a triangle on the stack holds the lengths of its sides as well as its vertices. If $n = \lceil \log(c/\sqrt{2}\delta) \rceil$, then the stack may be set at size $4n$.

15.6 CODE

The triangulation code, which includes the monotone partition, triangulation, and subdivision of triangles, is in `triangulation.inc`. Here in summary is the whole sequence of steps:

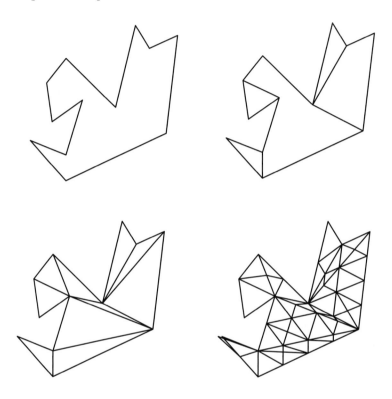

The procedure `monotone` has a single argument, an array of 2D points making up a closed polygon, and returns an array of polygons making up a monotone decomposition. The procedure `monotone-triangulation` has a single argument, a monotone polygon, and returns an array of triangles making up a triangular partition of it. The procedure `subdivide` has two arguments, a triangle and a real number δ, and returns an array of triangles partitioning it into triangles of diameter no more than δ. The procedure `triangulation` is the composite of these; it has two arguments, a polygon and δ, and returns an array of triangles of diameter at most δ.

The code for stacks is straightforward, implementing a stack as an array of length specified in advance, and can be found in `stack.inc`. This is a very simple package and can be used surprisingly often, if only as an expandable array. The procedures in this file are

Arguments	Command	Left on stack; side effects
n	`new-stack`	a stack of potential size *n*
x stack	`stack-push`	pushes *x* onto the stack
stack	`stack-peek`	returns the current stack top, leaving it in place
stack	`stack-pop`	removes and turns the current stack top
stack	`stack-length`	returns current size
stack	`stack-empty`	empty or not?
stack	`stack-full`	more room?
stack	`stack-extend`	doubles the size
stack	`stack-array`	returns an array of exactly the right size
n	`stack-element-at`	returns *n*th item from bottom on the stack

No checks are made on arrays when pushing or popping, and so underflow and overflow can occur, producing one of those dreaded PostScript error messages. But you can check for these possibilities with `stack-empty` and `stack-full`. Of course stacks should be very familiar to you by now, since PostScript uses them all the time.

The code for the binary search tree I use in triangulation is in `dynamic-tree.inc`. Code for a simpler binary search tree, the one discussed explicitly in Sedgewick's book, can be found in `simple-tree.inc`. Here the keys are integers indexing data to be stored in and retrieved from the tree. Sedgewick's book includes explicit code that turns out not to be too hard to render in PostScript. After all, algorithms are based on ideas more or less independently of language details. One major difficulty, and it indeed leads often to pain, is that in PostScript the only structures are arrays, and the only way to access items in an array is by index. Thus, whereas in C you can write `x->key = 3`, where x is a structure with a field named `key`, in PostScript you have to let x be an array and allocate one of its items, that with index 2 for instance, to a field you can think of as its key. Thus, `x->key = 3` in C becomes `x 2 3 put` in PostScript. Alas. Things like this can be done easily in principle, but in practice it's hard on the human memory to program like this. One small thing you can do, and I recommend it, is to make up a short dictionary of fields, here for example one defining KEY to be 2, so you can write `x KEY 3 put`. If you do this, you will want to `begin` this dictionary only when you want to use it and `end` it when you don't need it any more, or many different local conventions like this will get confounded.

Another thing to keep in mind when translating C to PostScript is that an array in either language, or a structure in C, is actually a **pointer** – that is, a single machine address in which the items in the array or structure are stored. Changing data in the structure does not change this address itself. Copying it does not give a new copy of the data in the structure but just a new copy of the address.

REFERENCES

1. M. de Berg, M. van Kreveld, M. Overmars, and O. Schwarzkopf, **Computational Geometry – Algorithms and Applications**, Springer-Verlag, 1997. The history of polygon triangulation is discussed at the end of Chapter 3. Planar sweeps are a commonly used technique in advanced graphics, and other examples of this technique can also be found here. Particularly interesting mathematically is the treatment of Voronoi cell construction in Chapter 7. The problem I look at in this chapter is typical of a huge family of similar graphics problems requiring relatively subtle techniques for their solution. This book has a good selection of further references.

2. Robert Sedgewick, **Algorithms in C**, Addison-Wesley, 1990. This is a deservedly popular text covering many basic computer algorithms. There is explicit code in C, but it can be read fruitfully without knowing much about that programming language. The illustrations accompanying the explanations of algorithms are invaluable. The most recent edition is **Algorithms in Java**, and in addition to a change of programming language there is also a much expanded exposition.

3. There is much information available on the Internet about computational graphics. Much of it can be used profitably in practical illustration problems. A good place to begin is Godfried Toussaint's Web site at

   `http://www-cgrl.cs.mcgill.ca/~godfried/teaching/cg-web.html`

APPENDIX 1

Summary of PostScript commands

This appendix offers a summary of PostScript operators useful for producing mathematical figures. Most have already been introduced. In addition, a few that are likely to be more rarely used than the rest are explained here. This is a large list, but by no means a complete one, of PostScript commands. The PostScript reference manual ("Red Book") contains a complete list by function as well as a list in alphabetical order in which the operators are described in occasionally invaluable detail.

There are many operators even in this restricted list, but fortunately most commands are very close to normal English usage and should be easy to remember.

The symbol $\emptyset$ means no arguments or no return value.

A1.1   MATHEMATICAL FUNCTIONS

Arguments	Command	Left on stack; side effects
$x\ y$	add	$x + y$
$x\ y$	sub	$x - y$
$x\ y$	mul	xy
$x\ y$	div	x/y
$x\ y$	idiv	the integral part of x/y
$x\ y$	mod	the remainder of x after division by y
x	abs	the absolute value of x
x	neg	$-x$
x	ceiling	the integer just above x
x	floor	the integer just below x
x	round	x rounded to the nearest integer
x	truncate	x with the fractional part chopped off
x	sqrt	square root of x
$y\ x$	atan	the polar argument of the point (x, y)
x	cos	$\cos x$ (x in degrees)

x	sin	$\sin x$ (x in degrees)
$x\ y$	exp	x^y
x	ln	$\ln x$
x	log	$\log x$ (base 10)
	rand	a random number

PostScript works with two kinds of numbers: integers and real. Real numbers are floating point with a limited number of decimals of accuracy. Arguments for some operations, such as repeat, must be integers. I leave as an exercise to tell whether ceiling etc. return – that is, leave on the stack – integers or real numbers. Many operations have an implicit range restriction; for example, sqrt must be applied to a nonnegative number.

A1.2 STACK OPERATIONS

x	pop	$\emptyset$
$x\ y$	exch	$y\ x$
x	dup	$x\ x$
$x_{n-1} \ldots x_0\ n\ i$	roll	$x_{i-1} \ldots x_0\ x_{n-1} \ldots x_i$

This rolls the top n elements on the stack around by a shift up of i elements. For example, if the stack holds 1 2 3 4 5 (from the bottom up), then 5 2 roll changes it to 4 5 1 2 3. It is more efficient if more complicated to do stack operations than access them by variable names, although the extra efficiency is often not worth the inconvenience of having to keep track of what's what on the stack.

$x_{n-1} \ldots x_0\ n$	copy	$x_{n-1} \ldots x_0\ x_{n-1} \ldots x_0$

A good trick for debugging is to combine copy and roll to view in a terminal window the top n items on the stack. The best way to do this (where $n = 3$) is

```
3 copy [ 4 1 roll ] ==
```

$x_i \ldots x_0\ i$	index	$x_i \ldots x_0\ x_i$

A1.3 ARRAYS

	[	begins an array
	]	closes an array
an array a	length	number of items in the array a
$a\ i$	get	a_i
$a\ i\ x$	put	$\emptyset$

Sets the ith entry of a equal to x. The way to remember the order of the arguments here is to think of this as formally equivalent to a[i] x def.

$a\ i\ j$	getinterval	$a_i \ldots a_j$
n	array	an empty array of length n with null entries

The null item in PostScript is like nothing ... else.

a	aload	$a_0 \ldots a_{\ell-1}\ a$ (ℓ is the length of a)

This essentially just unpacks a onto the stack but also puts a itself on top. If you want just to unpack a, use the pair aload pop.

An array in PostScript is what in other languages is called a **pointer**, which means it is stored in PostScript as an address in the machine in which the items in the array are stored. The practical importance of this is that if a is an array, then the sequence a dup doesn't make a new copy of the data stored by a but only a copy of the address where the data are stored. The sequence

```
a [ exch aload pop ]
```

will make a new array with the same data as a.

A1.4 DICTIONARIES

name item	def	makes an entry in the current dictionary
n	dict	puts a dictionary of n null entries on the stack
dictionary d	begin	opens d for use
	end	closes the last dictionary opened

Dictionaries in PostScript keep track of variable names and their current values. There may be several dictionaries in use at any moment; they are stored on a stack (the dictionary stack) and searched from the top down. The command begin puts a dictionary on this stack and end pops it off. So begin and end should be nested pairs.

something	bind	used before def to construct a procedure immediately

Normally, when defining a procedure, the names occurring in it are left as strings without attempting to look up their values when the definition is made. These names are looked up when the procedure is called. But when *bind* is used, the names that do occur in dictionaries are evaluated immediately.

A1.5   CONDITIONALS

The first few return "Boolean" constants true or false. A few others have Boolean values as arguments.

	false	false	(Boolean constant)
	true	true	(Boolean constant)
$x\ y$	eq	$x = y?$	
$x\ y$	ne	$x \neq y?$	
$x\ y$	ge	$x \geq y?$	
$x\ y$	gt	$x > y?$	
$x\ y$	le	$x \leq y?$	
$x\ y$	lt	$x < y?$	
$s\ t$	and	s and t are both true?	
$s\ t$	or	at least one of s and t is true?	
s	not	s is not true?	
$s \{\dots\}$	if	executes the procedure if s is true	
$s \{\dots\}\{\dots\}$	ifelse	executes the first procedure if s is true; otherwise, the second	

A1.6   LOOPS

$i\ h\ f\ \{\dots\}$	for	steps through the loop from i to f, incrementing by h

The tricky part of this is that at the start of each loop it leaves the loop variables i, $i + h, i + 2h$ on the stack. It is safest to use this only with integer loop variables.

$n \{\dots\}$	repeat	executes the procedure n times
$\{\dots\}$	loop	executes the procedure until exit is called from within the procedure
$\emptyset$	exit	exits the loop it is contained in
$\emptyset$	quit	stops everything
$a \{\dots\}$	forall	loops through the elements of a, leaving each in turn on the stack
{.}{.}{.}{.}	pathforall	loops through the current path (see immediately below)

The four arguments to pathforall are procedures to be called in the course of looking at the current path. This is a tricky command, but it can produce spectacular effects. A path is a special kind of array. Each element in it is one of the four commands x y moveto, x y lineto, x[1] y[1] x[2] y[2] x[3] y[3] curveto, closepath. The data are expressed in device coordinates. The command path-forall loops through the elements of the current path, pushing its arguments on the stack and then executing the corresponding procedure. For example, the following segment displays the current path:

```
{ [ 3 1 roll (moveto) ] == }
{ [ 3 1 roll (lineto) ] == }
{ [ 7 1 roll (curveto) ] == }
{ [ (closepath) ] == }
pathforall
```

The values of the coordinates are in the current user coordinates.

A1.7 CONVERSIONS

x s	cvs	an initial substring of the string *s* expressing *x*
x	cvi	*x* converted to integer

A1.8 FILE HANDLING AND MISCELLANEOUS

a string *s*	run	executes the file *s*
	showpage	changes a page
a procedure	exec	executes a procedure
a name	load	loads the value associated to the name

This is the way to put a named procedure on the stack without executing it.

Ø	save	puts a copy of the entire current state on the stack
state	restore	restores the state on the stack

Thus,

```
save /SavedState exch def
...
SavedState restore
```

will save and restore a snapshot of a state.

	type	tells what type the object at the top of the stack is

It pops that object from the stack, and so you will likely want to use dup and type together. This is one of the more complicated PostScript operators. First of all, what it returns is one of the following names:

arraytype	an array
booleantype	a boolean like true or false
dicttype	a dictionary
fonttype	a font
integertype	an integer like 1
marktype	a [
nametype	a name like/x
nulltype	a null object
operatortype	an operator like add
realtype	a real number like 3.14159
stringtype	a string like (x)

or possibly one of a few types I haven't introduced.

Second, what it returns is an executable object, which means if you apply to it the exec operator it will execute whatever has been defined by you to be associated to that name. Thus, after

```
/arraytype { dup length = == } def
/integertype { = } def
```

the sequence dup type exec will display and pop the object at the top of the stack if it is an integer, display and pop it and its length if it is an array, and give you an undefined error otherwise. This allows you to have a procedure do different things, depending on what kind of arguments you are passing to it. The PostScript operator transform behaves like this, for example, detecting whether the top of the stack contains a matrix or a number.

A1.9   DISPLAY

x	=	pops x from the stack and displays it on the terminal
x	==	almost the same as =

The most important difference between the two is that the operator == displays the contents of arrays, whereas = does not. One curious difference is how they handle strings. Thus (x) = displays x in the terminal window, whereas (x) == displays (x). In particular, it is useful when using terminal output for debugging to know that () = produces an empty line.

...	stack	displays the whole stack (but not arrays), not changing it
...	pstack	same as stack. but also displays arrays
string s	print	prints a string; has better format control than the others

The difference between = and == is that == will display arrays and = will not. Sometimes this is a good thing, and sometimes not; sometimes arrays will be huge and displaying them will fill up your screen with garbage. The difference between stack and pstack is the same.

As for print, it is a much fancier way to display items – more difficult to use but with output under better control. For example,

```
(x = ) print
x (      ) cvs print
(\n) print
```

will display "x =" plus the current value of *x* on a single line. What's tricky is that print displays only strings, and so everything has to be converted to one first. That's what cvs does. The (\n) is a string made up of a single carriage return because otherwise print doesn't put one in.

A1.10 GRAPHICS STATE

| Ø | gsave | saves the current graphics state and installs a new copy of it |
| Ø | grestore | brings back the last graphics state saved |

The graphics state holds data such as the current path, current linewidth, current point, current color, current font, and so on. These data are held on the **graphics stack**, and gsave and grestore put stuff on this stack and then remove it. They should always occur in nested pairs. All changes to the graphics state inside a pair have no effect outside it. It is a good idea to encapsulate all fragments of a PostScript program that change the graphics state to draw something inside a gsave ... grestore pair unless you really want a long-lasting change.

We have seen three stacks used by a PostScript interpreter: the **operator stack**, which is used for calculations; the **dictionary stack**, which controls access to variable names; and the **graphics stack**. There is one other stack, the **execution stack**, which is used to keep track of what procedures are currently running; however, the user has little explicit control over it, and it is not important to know about it.

x	setlinewidth	sets current linewidth to x (in current units)
	currentlinewidth	the current linewidth in current units
x	setlinecap	determines how lines are capped
x	setlinejoin	determines how lines are joined
$[\ldots]\, x$	setdash	sets current dash pattern

For example [3 2] 1 `setdash` makes it a sequence of dashes 3 units long and blanks 2 units long each with an offset of 1 unit at the beginning.

Experimentation with `setdash` can be interesting. The initial array specifying the on–off pattern can be long and complicated and itself produced by a program. Go figure.

g	`setgray`	sets current color to a shade of gray
$r\ g\ b$	`setrgbcolor`	sets current color

In both of these, the arguments should be in the range $[0, 1]$.

A1.11 COORDINATES

Here, a matrix is an array of six numbers. The CTM is the **C**urrent **T**ransformation **M**atrix.

$\emptyset$	`matrix`	puts a matrix on the stack
matrix m	`defaultmatrix`	fills m with the default TM; leaves it on the stack
m	`currentmatrix`	fills the matrix with the current CTM; leaves it
$x\ y$	`translate`	translates the origin by $[x, y]$
$a\ b$	`scale`	scales x by a, y by b
A	`rotate`	rotates by Ac degrees
m	`concat`	multiplies the CTM by m
m	`setmatrix`	sets the current CTM to m
m	`identmatrix`	puts the identity matrix inter the matrix m
$x\ y$	`transform`	$x'\ y'$, transform of $x\ y$ by the CTM
$x\ y\ m$	`transform`	$x'\ y'$, transform of $x\ y$ by m
$x\ y$	`itransform`	$x'\ y'$, transform of $x\ y$ by the inverse of the CTM
$x\ y\ m$	`itransform`	$x'\ y'$, transform of $x\ y$ by the inverse of m

There are also operators `dtransform` and `idtransform` that apply just the linear component of the matrices (to get relative position).

$m_1\ m_2$	`invertmatrix`	m_2 (the matrix m_2 is filled by the inverse of m_1)

A1.12 DRAWING

$\emptyset$	`newpath`	starts a new path, deleting the old one
$\emptyset$	`currentpoint`	the current point $x\ y$ in device coordinates

In order for there to be a current point, a current path must have been started. Every path must begin with a `moveto`, and so an error message complaining that there is no current point probably means you forgot a `moveto`.

$x\ y$	`moveto`	begins a new piece of the current path
$x\ y$	`lineto`	adds a line to the current path
$dx\ dy$	`rmoveto`	relative move
$dx\ dy$	`rlineto`	relative line
$x\ y\ r\ a\ b$	`arc`	adds an arc from angle a to angle b, center (x, y), radius r
$x\ y\ r\ a\ b$	`arcn`	negative direction

The operator `arc` is rather complicated. If there is no current path under construction, it starts off at the first angle and makes the arc to the second. If there is a current path already, it adds to a line from where it ends to the beginning of the arc before it adds the arc to the current path. Similarly for `arcn`.

$x_1\ y_1\ x_2\ y_2\ x_3\ y_3$	`curveto`	adds a Bezier curve to the current path
$dx_1\ dy_1\ dx_2\ dy_2\ dx_3\ dy_3$	`rcurveto`	coordinates relative to the current point
Ø	`closepath`	closes up the current path back to the last point moved to
Ø	`stroke`	draws the current path
Ø	`fill`	fills the outline made by the current path
Ø	`clip`	clips drawing to the region outlined by the current path
Ø	`pathbbox`	$x_\ell\ y_\ell\ x_u\ y_u$

This returns four numbers `llx lly urx ury` on the stack that specify the lower left and upper right corners of a rectangle just containing the current path.

Ø	`strokepath`	replaces the current path by its outline
a special dictionary	`shfill`	used for gradient fill

A1.13 DISPLAYING TEXT

font name	`findfont`	puts the font on the stack
font s	`scalefont`	sets the size of the font (in current units) and leaves it on the stack
font	`setfont`	sets that font to be the current font

So that

```
/Helvetica-Bold findfont
12 scalefont
setfont
```

sets the current font equal to Helvetica–Bold at an approximate height of 12 units.

string *s*   show   displays *s*

The string is placed at the current point and moves that current point to the end of the string. Usually it is prefaced by a moveto. There must also be a current font set.

string *s*   stringwidth w_x w_y, the shift caused by showing *s*

That is, displaying a string moves the current point. This returns the shift in that point.

string *s* boolean *t*   charpath   the path this string would make if displayed.

Use true for filling or clipping the path and false for stroking it. In some circumstances these will produce somewhat different results, and in particular the path produced by true might not be what you want to see stroked.

A1.14 ERRORS

When a program encounters an error, it displays a key word describing the type of error it has met. Here are some of the more likely ones, roughly in the order of frequency, along with some typical situations that will cause them.

undefined	A word has been used that is undefined. Often a typing error.
rangecheck	An attempt has been made to apply an operation to something not in its range.

For example, -1 sqrt or [0 1] 2 get.

syntaxerror	Probably an (or { without matching) or }.
typecheck	An attempt to perform an operation on an unsuitable type of datum.
undefinedfilename	An attempt to run a file that doesn't exist.
undefinedresult	5 0 div
unmatchedmark	] without a previous [.
dictstackoverflow	Dictionaries have not been closed. Probably a begin without end.

A1.15 ALPHABETICAL LIST

Here is a list of all the operators described above along with the section each can be found in.

=	9	end	4
==	9	eq	5
[	4	exch	2
]	4	exec	8
abs	1	exit	6
add	1	exp	1
aload	3	false	5
and	5	fill	12
arc	12	findfont	13
arcn	12	floor	1
array	3	for	6
atan	1	forall	6
begin	4	ge	5
bind	4	get	3
ceiling	1	getinterval	3
charpath	13	grestore	10
clip	12	gsave	10
closepath	12	gt	5
concat	11	identmatrix	11
concatmatrix	11	idiv	1
copy	2	idtransform	11
cos	1	if	5
currentlinewidth	10	ifelse	5
currentmatrix	11	index	2
currentpoint	12	invertmatrix	11
curveto	12	itransform	11
cvi	6	le	5
cvs	6	length	3
def	4	lineto	12
defaultmatrix	11	ln	1
dict	4	load	8
dictstackoverflow	14	log	1
div	1	loop	6
dtransform	11	lt	5
dup	2	matrix	11

mod	1	scalefont	13
moveto	12	setdash	10
mul	1	setfont	13
ne	5	setgray	10
neg	1	setlinecap	10
newpath	12	setlinejoin	10
not	5	setlinewidth	10
or	5	setmatrix	11
pathforall	6	setrgbcolor	10
pathbbox	12	shfill	12
pop	2	show	13
print	9	showpage	8
pstack	9	sin	1
put	3	sqrt	1
quit	6	stack	9
rand	1	stringwidth	13
rangecheck	14	stroke	12
rcurveto	12	strokepath	12
repeat	6	sub	1
restore	8	syntaxerror	14
rlineto	12	transform	11
rmoveto	12	translate	11
roll	2	true	5
rotate	11	truncate	1
round	1	typecheck	14
run	7	undefined	14
save	8	undefinedfilename	14
scale	11	undefinedresult	14

APPENDIX 2

Setting up your PostScript environment

To run PostScript programs, you will need to have a PostScript viewer installed on your machine. The most convenient way to do this is to install the basic PostScript interpreter Ghostscript and then on top of that one of several possible interactive viewers that call on Ghostscript for basic graphics rendering. The program Ghostscript is available without cost for download from `http://www.cs.wisc.edu/~ghost/`. The viewers `GhostView`, `GSView`, `MacGSView`, and `GV` (for various platforms) can also be found there.

On Unix and Macs the command line interface for the interpreter Ghostscript (as opposed to a file viewer) should be straightforward to figure out, but for Windows machines it is a little more difficult. First run (i. e., **Run**) the program `cmd.exe`, and then in the terminal window that pops up type `gswin32c.exe` togther with various options to get Ghostscript on its own. One variant that you can use fruitfully for debugging is

```
gswin32c.exe -dNODISPLAY <filename>
```

You should be able to set shortcuts up so that not so much typing is involved.

A2.1 EDITING PostScript FILES

It is important to use the right text editor in writing PostScript programs or at least to know how to use correctly the one that you do use. First of all, a PostScript file must be just an ordinary text file without formatting adornments such as those produced by Microsoft **Word** in its default configuration. So you must be careful, if necessary, to save your file as a plain text file. In some text editors, text files will be automatically given an extension `.txt`. This is not necessarily a problem, but for your own long-term sanity it is probably best to store all your PostScript files with

271

an extension .ps (or a variation like .eps). This may require that you explicitly rename each file.

A2.2   RUNNING EXTERNAL FILES

Once you have installed Ghostscript and a viewer, you will have to do a little work to configure your environment for easy PostScript program development. I have described in this book a number of packages of PostScript procedures that you will want to incorporate in your own programs with the PostScript run command. This command simply loads a file that it interprets as any other sequence of PostScript code. But for security reasons, the way most PS viewers are configured by default is to disallow this. To allow it you must turn off the viewer option usually associated with the keyword **Safer**. In GSView, for example, you can do this by opening the **Options** menu (here showing **Safer** toggled on): If you do this and you use your viewer to look at PostScript files on the Internet from within your browser, you should be sure to make the **Safer** option still in force inside the browser. How to do this depends on which browser and which operating system you are using. With my viewer gv inside Netscape, I set the application that reads PostScript documents to be gv -safer %s.

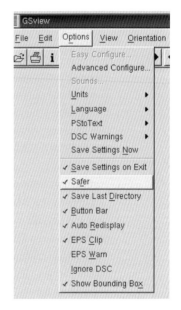

Another remark about the run command: recent versions of Ghostscript (8.0 and after) set some files (those with an EPSF in the first line) to be run differently from others. In effect, alas, the default Ghostscript is no longer a strict PostScript interpreter since it reads comments. This dubious "feature" should probably be normally disabled by specifying the option gs -dNOEPS in your viewer.

A2.3   MAKING IMAGES

At some point you will probably want to make image files from PostScript programs such as .jpg or .gif images that you can include directly inside a Web page. There are several ways to do this, one being to use Ghostscript itself to do the job, or your Ghostscript viewer. With GSView, for example, you can see what your options are if you look at the menu item File/Convert. But you will have somewhat more control over the output if you use a full-featured image manipulation program.

The most comprehensive of these is *PhotoShop*, but it is expensive and complicated. Another possibility is the program **GIMP** (**G**nu **I**mage **M**anipulation **P**rogram), available without cost for many platforms.

■ *To use GIMP on Windows machines to interpret PostScript programs, it must be able to locate the Ghostscript executable file. The simplest way to do this is to copy the file* gswin32c.exe, *which is put somewhere on your computer as part of the installation of Ghostscript, into the directory* C:\Windows.

One thing to be careful about when importing a PostScript file into an image manipulation program is that many of them require that the file be terminated with showpage. Another is that all files that are imported into your file with a run command must normally be included in the file explicitly. I'll discuss this problem in the next section.

When you import a PostScript file into an image manipulation program, you will likely have some choices to make about the size and quality of the import. The most subtle choice to make is about **antialiasing**. A PostScript file is normally scalable, which means it has no intrinsic granularity. But an image manipulation program will turn this into a bit map, which is an image with a particular resolution. In other words, it works with discrete units of color called pixels. In rendering a scalable picture into a bit map, different algorithms are possible. The highest quality will be obtained if transitions of pixel colors are as smooth as possible, and this is what antialiasing accomplishes. The term "aliasing" here comes from the theory of the Fourier transform, but to explain how it applies would take up far too much space.

A2.4 PRINTING FILES

You can print PostScript files if you have a PostScript driver for your printer. This is no problem on some operating systems, but on Windows machines it is not present in the usual installation. It will be possible once you have Ghostscript installed, however. You can also use your PostScript viewer to print PostScript files with a few mouse clicks.

In any event, there are a few precautions to observe. First of all, most printers will not print a page unless it is terminated by showpage. Second, the printer will not be able to load files referred to with a run command, and so you must actually include such files inside your program before printing them. (This is usually true of files imported into an image manipulation program also.) Your text editor almost certainly allows file inclusion, but this is a relatively slow process and is awkward because you will usually be much happier doing your own editing in a file without the package included. Once you include the file explicitly, then, your own program development will be somewhat restricted. I prefer to make the process somewhat automatic as well as flexible. For this purpose I do the following: (1) I write my original program in a file called x.px for instance. It may contain several run commands. (2) I apply a PERL script I call psinc, which includes explicitly all the files referred to by run commands and then creates a new file x.ps, which is self-contained. Furthermore, it includes files recursively, which means that even the files that are included may themselves run other files. I work with a Linux system, but this sort of thing should be possible on almost any system. Here is my PERL script:

```perl
#!/usr/bin/perl

# This reads in ps files, printing out all lines except
# ^( ... ) run
# where it performs an inclusion --- recursively.
# The current directory is updated.
use File::Basename;
include("", STDIN, 0, "stdin");

# insert a file into output
sub include {
    local($curdir, $input, $depth, $cf) = }@_;
    $i = $depth;
    while ($i > 0) {
        print STDERR "   ";
        $i--;
    }
    print STDERR "Opening $cf\n";
    $fh++;
    while ($_ = <$input>) {
        if (/^\ ((.*)\)[ ]*run/) {
            ($name, $dir, $suffix) = fileparse($1, ");
            if ($dir eq './') {
                $dir = "";
            }
            $file = "$curdir$dir$name$suffix";
            # print STDERR "file to open = {$file}\n";
            if (open($fh, $file)) {
                print "\n";
                print "% - Inserting $name ------------------\n\n";
                include($curdir.$dir, $fh, $depth+1, $file);
            } else {
                print("Current directory $curdir.$dir\n");
                print STDERR "Unable to open $file\n";
                exit 1;
```

```
            }
        } else {
            print $_;
        }
    }
    if ($input ne STDIN) {
        $i = $depth;
        while ($ i > 0) {
            print STDERR "    ";
            $i--;
        }
        print "\n";
        print "% - closing $name ----------------------\n";
        print STDERR "Closing $cf\n";
        close($input);
    }
}
```

On my Linux system, I have available to me the make utility, and turning a .px file
into a .ps file is an option built into my make configuration:

```
.SUFFIXES: .px .ps


.px.ps:
        rm -f $*.ps; EPS2eps < $*.px } psinc > $*.ps
```

The whole process therefore becomes quite painless. I just type make x.ps
when x.px.ps is changed.

Structured PostScript documents

A PostScript program is just a sequence of PostScript commands to be interpreted in the order in which they are encountered. It swallows one command after another. Once it has executed a sequence of commands, it essentially forgets them. In short, the PostScript interpreter knows nothing about the global structure of your file and in particular has no idea of the separate pages as individual items.

But there are conventions that allow you to put such a structure in your file. These are called **document structure comments**. You will likely have seen these if you ever peeked at a PostScript file produced by some other program. For example, the program dvips that turns the .dvi files produced by the mathematical typesetting program TEX into PostScript might produce a file that looks like this:

```
%!PS-Adobe-2.0
%%Creator: dvips(k) 5.86 Copyright 1999 Radical Eye Software
%%Title: a3.dvi
%%Pages: 2
%%PageOrder: Ascend
%%BoundingBox: 0 0 596 842
%%DocumentFonts: Helvetica-Bold Palatino-Roman CMTT10
%%EndComments
%DVIPSWebPage: (www.radicaleye.com)
%DVIPSCommandLine: dvips a3.dvi -o a3.ps
%%BeginProcSet: texc.pro

...

%%EndProlog
%%BeginSetup
%%Feature: *Resolution 600dpi
```

```
TeXDict begin
%%PaperSize: letter
%%EndSetup
%%Page: 1 1
1 0 bop 0 191 a Fd(Appendix)24 b(3.)36 b(Structured)24
b(P)l(ostScript)i(documents)0 540 y Fc(A)h(PostScript)h(pr)o(ogram)f
...
%%Trailer
end
userdict /end-hook known{end-hook}if
%%EOF
```

Indeed, this is the PostScript file making up this very appendix! All those lines start-ing with %% are **document structure comments** (with acronym **DSC**) meant to be interpreted by a program such as a PostScript viewer that allows it to output some information about the source of this document and permits a reader to move around in it from one page to the other but not necessarily in the order in which the pages were naturally encountered. Some of these comments are more important than the others – in particular those allowing moving around among the pages. And any PostScript program that intends to allow this sort of page-by-page viewing must follow certain conventions that make it feasible.

A PostScript program following DSC conventions should begin with a line such as

    %!PS-Adobe-2.0

that tell a viewing program that the appropriate conventions have been followed. The 2.0 refers to a version number for the conventions. I have to confess that in practice much slop is tolerated here and that I just put that very line at the beginning of almost all my own PostScript programs! This may very well cause the viewer to excrete nasty looking warning messages, but these can be ignored or even switched off by choosing options suitably.

Second, the page structure has to be indicated. Toward the beginning of the file there should be a line like

    %%Pages: 7

and as each page is begun there should be a line like

    %%Page: 1 1

The duplication of numbers here is not necessary but in handwritten PostScript code is probably the right thing to do (one number indicates the real page number, the other a nominal page number).

The leading line %!PS-Adobe-xxx and the %%Pages:x and %%Page:x x comments are the minimal set of comments needed to guide the viewer. But for the viewer not to be confused, it is extremely important to make the partition into pages meaningful:

■ *In a PostScript file setting up a page structure, each page must be independent of every other page so that pages may be interpreted correctly no matter the order in which they are read.*

This means above all that no variable may be defined on one page and used on another without an independent definition. As a general rule, all procedures to be used in the program should be put in a preliminary section of the file called the **prolog**. A coordinate system should be set up anew on each page and encapsulated within a gsave ... grestore pair (as I have recommended already in Chapter 1). Here is a rather simple PostScript program demonstrating these points:

```
%!PS-Adobe-2.0
%%Pages: 2
/page-begin {
  gsave
  72 dup scale
  1 72 div setlinewidth
} def
/page-end {
  grestore
  showpage
} def
%%Page: 1 1
page-begin
...
% draw something
...
page-end
%%Page: 2 2
page-begin
...
% draw something else
...
page-end
```

APPENDIX 4

Simple text display

You will very often want to put text in your figures. PostScript's font-handling capabilities are extremely good, but most of the techniques for high-quality font management are designed to be automated by some other program because good text – especially mathematical text – requires much computation to get font choices, spacing, and sizes right.

In this appendix I'll explain how to place simple text in PostScript figures and also a few playful possibilities. What I explain here will be adequate for many purposes. (See Appendix 7 for more sophisticated techniques.)

A4.1 SIMPLE PostScript TEXT

The simplest, essentially the only simple, way to put text into pictures is to use the almost universally available PostScript fonts to assemble your text "by hand," that is, by thinking out for yourself what layout, font choice, and text size are to be. This is a relatively straightforward process and is probably your best choice if the text you want to include is not too complicated.

There are three steps to be carried out each time you want to use a new font.

(1) You must decide which font you are going to use. There is a limited choice of fonts guaranteed to be available in all environments. The choice is, roughly, from this list:

Times-Roman	Times-Italic	Times-Bold	Times-Bolditalic
Helvetica	Helvetica-Oblique	Helvetica-Bold	Helvetica-Bold-Oblique
Courier	Courier-Oblique	Courier-Bold	Courier-Bold-Oblique
Symbol			

The Helvetica fonts have no serifs and display well on a computer. The Courier fonts have uniform character spacing. The Symbol font contains Greek letters.

You load a font with the command `findfont` applied to the name of the font.

(2) You decide what scale you are going to use it at. With the command `scalefont`, you set what is essentially the vertical size of the letters in terms of the current unit.

(3) You apply the command `setfont`.

If the current unit is 1 inch, for example, this will give you letters 1/4″ high:

```
/Helvetica-Bold findfont
0.25 scalefont
setfont
```

The command `findfont` loads the font named onto the stack. The command `scalefont` sets the size of the font on the stack, leaving the font there. The command `setfont` sets the current font to the one on the top of the stack. The current font is part of the graphics state, and thus it is affected by the `gsave` and `grestore` commands.

Of course, you can switch back and forth among several fonts. If you are going to do this, you will probably want to write procedures to do this efficiently rather than having to go through the whole sequence just described.

You can use other PostScript fonts, too, but you ought to include them explicitly in your file.

To put text on a page after you have set a current font, move to where you want the text to begin and then use the `show` command. The text itself is made into a **string** by enclosing it within parentheses. Thus,

```
0 0 moveto
(Geometry) show
```

after the previous command sequence will give you

Geometry

You can print out the value of a variable on the page. To do this, you must convert the variable to a string with the command `cvs` using an empty string of sufficient size to hold the variable's value.

```
0 0 moveto
(x = ) show
x (        ) cvs show
```

will produce this, if $x = 3$:

x = 3

A font is part of the graphics state, and the way it's displayed depends entirely on the current coordinate system. You can understand exactly what happens if you keep this in mind:

■ *PostScript treats letters as paths.*

Scaling of a font is in terms of the current units. Changing the scale of the entire figure will also affect a font's true size along with the size of everything else. But you can also shear a font or reverse it by suitable `concat` operations. You can get interesting effects.

Geometry

A4.2   OUTLINE FONTS

If characters are paths, you should be able to do with them all the things you can do with ordinary paths. This is almost true; the only exception is that often the inner details of fonts are hidden from close inspection. This is for legal reasons. Characters from a font, along with all other images, are subject to copyright, but because of their high reproducibility they are often encrypted. On the other hand, many high-quality fonts are in the public domain and unencrypted.

The command `show` applied to a string fills in a special way the path generated by the string. But you can access the path itself by using the command `charpath`, which appends the path of the string to the current path. You could then fill it, but for that task this wouldn't be a very efficient way of proceeding. One more interesting thing you can do, however, is stroke it or clip to it. The command `charpath` takes two arguments, a string and a Boolean value. Use `true` if you intend to stroke the outline and `false` if you intend to fill or clip. It is also a good idea to set the values of `linejoin` and `linecap` to something other than 0.

```
1 setlinejoin
1 setlinecap

/Helvetica-Bold findfont
48 scalefont
setfont
```

```
gsave
newpath
0 0 moveto
(ABC) false charpath
clip
1 0 0 setrgbcolor
newpath
2 setlinewidth
-24 8 96 {
  /i exch def
  i 0 moveto
  40 40 rlineto
} for
stroke
grestore

newpath
0 0 moveto
(ABC) true charpath
0.5 setlinewidth
stroke
```

APPENDIX 5

Zooming

One of the greatest advantages of using PostScript for illustrations is that it is scalable, and thus there are no artifacts in the illustration that show up when it is examined closely. This is in opposition to digital photographs, for example, when blowup will start to show pixels. This appendix will explain how to take advantage of this.

A5.1 ZOOMING

I will explain here a procedure called `zoom` that has the effect of zooming in at a point by a given scale. The overall effect can be illustrated by these three figures, where the zoom factor is 2:

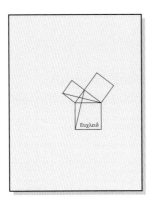 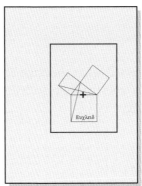 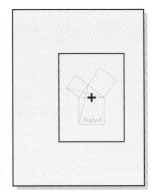

How can we do this? There are three arguments for this procedure. One is the scale factor c. If it's greater than 1, the scale change is a magnification and we are zooming in. If it's less than 1, we are zooming out, not in. If it's exactly 1, there is no scale change; the zoom will amount to a translation of the origin. Another argument is a point (x, y) in the original figure. The last argument is the point (c_x, c_y) to which (x, y) is to be relocated. If we want to locate (x, y) at the center of a page, for example, and if the current coordinate system is the page coordinate system, then $(c_x, c_y) = (306, 396)$. But if the origin of the current coordinate system is already at the center of the page, it is $(0, 0)$.

I call c the **zoom factor**, (x, y) the **focus** of the zoom, and (c_x, c_y) its **center**.

It is more or less clear that what we want is a succession of translation and scales, but in what order? And which ones? The simplest way to decide is to portray geometrically what has to be done:

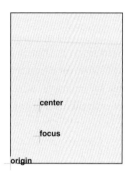  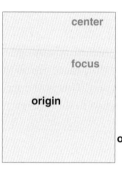 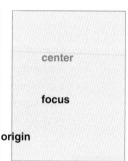

This leads to the following code:

```
cx cy translate
s dup scale
x neg y neg translate
```

which is to be inserted before the original drawing commands.

If you want to rotate the figure with the focus as the pivot, then add the correct line as is done here:

```
cx cy translate
20 rotate
s dup scale

x neg y neg translate
```

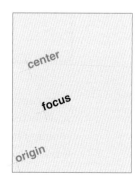

A5.2 AN EXPLICIT PROCEDURE

Call the following procedure before drawing:

```
% On the stack when called are
% [cx cy] [x y] s: the place that is now (x, y) is located at [cx cy]
% and lengths scaled by s
/zoom { 3 dict begin
  /s exch def
  aload pop
  /y exch def
  /x exch def
  aload pop
  translate
  s dup scale
  x neg y neg translate
  currentlinewidth s div setlinewidth
end } def
```

A5.3 PLAYING AROUND

Try this:

```
(zoom.inc) run
/draw {
  gsave
  1 0 0 setrgbcolor
  newpath
  x y moveto
  -100 0 rlineto
   200 0 rlineto
  x y moveto
  0 -100 rlineto
  0 200 rlineto
  stroke
  grestore
  x y moveto
  (Euclid) show
} def

/Helvetica-Bold findfont
25 scalefont
setfont
```

```
/s 1 def
/x 100 def
/y 100 def
{  % loop
   gsave
   [x y] [x y] s zoom
   draw
   grestore
   /s s 1.1 mul def
   showpage
} loop
```

How would you get the text to rotate around the focus as the loop proceeds?

A5.4   CODE

See zoom.inc. There is a variant in there of the procedure zoom called Zoom. The third argument for this procedure is an array of four numbers specifying a linear transformation to be applied at the focus of the zoom. Used with [s 0 0 s], for example, it is equivalent to a zoom with scale factor *s*.

Evaluating polynomials: getting along without variables

Being able to evaluate arbitrary polynomials is very useful. We want a procedure with two arguments, the first a number x and the second an array $[a_0\ a_1 \ldots a_{n-1}]$ to be interpreted as the coefficients of a polynomial. The procedure should return $P(x) = a_0 + a_1 x + \cdots + a_{n-1} x^{n-1}$. In choosing this order for the arguments, I am following the usual rule of PostScript with an argument x first and then the object to be applied to it (the polynomial). The point is that this choice makes composition easy.

In using polynomial evaluation in some tools, such as mkpath, the derivatives of P are also needed. The method used to evaluate $P(x)$ can evaluate $P'(x)$ with little extra effort.

In many applications, a polynomial has to be evaluated many times, and it is therefore important to design the evaluation procedure to be efficient. This will offer an excuse to include a few remarks about managing the stack without variable names.

A6.1 THE MOST STRAIGHTFORWARD WAY TO DO IT

Here is a simple procedure that will evaluate an arbitrary polynomial $a_3 x^3 + a_2 x^2 + a_1 x + a_0$ of degree three.

```
% arguments: number x and array a = [ a0 a1 a2 a3 ]
cubic-poly { 2 dict begin
  /a exch def
  /x exch def
  a 0 get
  a 1 get x mul add
```

```
    a 2 get x 2 exp mul add
    a 3 get x 3 exp mul add
end } def
```

EXERCISE A6.1. *Extend this procedure using a* for *loop so that it will evaluate a polynomial of arbitrary degree. Be careful that your procedure works even for a polynomial of degree 0 (a constant). Also, it should return 0 if the array is empty. (Note that the degree is one less than the length of the coefficient array.)*

EXERCISE A6.2. *Extend in turn the procedure from the previous exercise so that it will return the array of two numbers* $[P(x)\ P'(x)]$.

A6.2 HORNER'S METHOD

The PostScript command exp is somewhat slow, and the straightforward procedure used above is therefore probably inefficient. Better is an elegant method of evaluating polynomials due to the nineteenth-century English mathematician W. G. Horner. It does not use exp but gets by with just successive multiplications and additions.

We start off by rewriting a cubic polynomial:

$$P(x) = a_3 x^3 + a_2 x^2 + a_1 x + a_0 = (((a_3)x + a_2)x + a_1)x + a_0 .$$

In other words, to evaluate the polynomial it suffices to calculate in succession

a_3

$a_3 x + a_2$

$(a_3 x + a_2)x + a_1$

$(((a_3)x + a_2)x + a_1)x + a_0$

or, in other words, giving these expressions labels, we calculate

$b_2 = a_3$

$b_1 = b_2 x + a_2$

$b_0 = b_1 x + a_1$

$b_{-1} = b_0 x + a_0 .$

At the end $b_{-1} = P(x)$. In PostScript this becomes

```
/b a 3 get def
/b b x mul a 2 get add def
/b b x mul a 1 get add def
/b b x mul a 0 get add def
```

and at the end $b = P(x)$. We can even get by without definitions.

```
a 3 get
x mul a 2 get add
x mul a 1 get add
x mul a 0 get add
```

will leave $P(x)$ on the stack. There is the germ of a simple loop here.

EXERCISE A6.3. *Write a procedure that evaluates an arbitrary fourth-degree polynomial $P(x)$ using a loop and without defining b.*

It would be easy enough to construct a PostScript procedure that implements Horner's algorithm using variables x and a, but it a bit more interesting to construct one that does all its work on the stack. The point of this is that accessing the value of a variable is an operation costly in time. Besides, it's often an enjoyable exercise.

I haven't said much about what's involved in sophisticated stack management. The most important thing to keep in mind is this:

■ *If you are going to get along without variable names, then the stack has to hold the entire state of the computation at every moment.*

In Horner's method, the state of the computation involves the value of x, the current value of the polynomial, and a specification of the coefficients yet to be used. In the following code, this is encapsulated in the list of unused coefficients a_i, the current polynomial value P, and the variable x sitting on the stack bottom to top in that order inside a repeat loop.

```
% x [ a0 a1 ... ]
/horner {
aload length              % x a0 a1 ... an n+1
dup 2 add -1 roll         % a0 a1 ... an n+1 x
exch 1 sub {              % a0 a1 ... P=an x
  dup 4 1 roll            % a0 ... x ak P x
```

```
      mul add exch              % a0 ... a[k-1] P x
    } repeat
    % at end P x on stack
  pop                           % P

} def
```

A6.3 EVALUATING THE DERIVATIVES EFFICIENTLY

Very often in plotting a graph it is useful to obtain the value of $P'(x)$ at the same time as $P(x)$. Horner's method allows this to be done with little extra work.

The formulas for the b_i above can be rewritten as

$$a_3 = b_2$$
$$a_2 = b_1 - b_2 x$$
$$a_1 = b_0 - b_1 x$$
$$a_0 = b_{-1} - b_0 x.$$

We can therefore write

$$
\begin{aligned}
P(X) &= a_3 X^3 + a_2 X^2 + a_1 X + a_0 \\
&= b_2 X^3 + (b_1 - b_2 x) X^2 + (b_0 - b_1 x) X + (b_{-1} - b_0 x) \\
&= b_2 (X^3 - X^2 x) + b_1 (X^2 - X x) + b_0 (X - x) \\
&= (b_2 X^2 + b_1 X + b_0)(X - x) + b_{-1} \\
&= (b_2 X^2 + b_1 X + b_0)(X - x) + P(x).
\end{aligned}
$$

Consequently, the coefficients b_i have significance in themselves; they are the coefficients of a simple polynomial:

$$b_2 X^2 + b_1 X + b_0 = \frac{P(X) - P(x)}{X - x} = \text{(say) } P_1(X).$$

One remarkable consequence of this is that we can evaluate $P'(x)$ easily because a simple limit argument gives

$$P'(x) = b_2 x^2 + b_1 x + b_0,$$

which means that we can apply Horner's method to the polynomial

$$P_1(X) = b_2 X^2 + b_1 X + b_0$$

in turn to find it. We can in fact evaluate $P(x)$ and $P'(x)$ more or less simultaneously if we think a bit about it. Let the numbers c_i be calculated from $P_1(X)$ in the same way that the b's came from $P(X) = P_0$:

$$b_2 = a_3$$
$$b_1 = b_2 x + a_2$$
$$b_0 = b_1 x + a_1$$
$$b_{-1} = b_0 x + a_0$$
$$c_1 = b_2$$
$$c_0 = c_1 x + b_1$$
$$c_{-1} = c_0 x + b_0$$

concluding with $P(x) = b_{-1}$, $P'(x) = c_{-1}$. This requires that we store the values of b_i to be used in calculating the c's. In fact we can avoid this by interlacing the calculations:

$$b_2 = a_3$$
$$c_1 = b_2$$
$$b_1 = b_2 x + a_2$$
$$c_0 = c_1 x + b_1$$
$$b_0 = b_1 x + a_1$$
$$c_{-1} = c_0 x + b_0$$
$$b_{-1} = b_0 x + a_0$$

concluding with $P(x) = b_{-1}$, $P'(x) = c_{-1}$.

EXERCISE A6.4. *Design a procedure with two arguments, an array and a number x, that returns $[P(x)\ P'(x)]$, where the polynomial P is defined by the array argument. It should use Horner's method to do this, preferably in the most efficient version. (Hint: Evaluating the derivative requires one less step than evaluating the polynomial. This can be dealt with by initializing c to 0 in the steps shown above. In this way, c and b can be handled in the same number of steps, and writing the loop becomes simpler.)*

A6.4 EVALUATING BERNSTEIN POLYNOMIALS

The Bernstein polynomials are the generalizations of the Bézier cubic polynomials, which are polynomials of the form

$$B_y(t) = (1-t)^n y_0 + n(1-t)^{n-1} t y_1 + \frac{n(n-1)}{2}(1-t)^{n-2} t^2 y_2 + \cdots + t^n y_n.$$

They are used frequently in computer graphics for reasons explained elsewhere. A procedure to evaluate one has two arguments, the number t and the array of the y_i, and returns $B_y(t)$. The basic principle here will be as in Horner's algorithm except that in this case the polynomial coefficients must be calculated as we go along. Fix n and let

$$C_k = \frac{n(n-1)\ldots(n-(k-1))}{k!} t^k$$

so that

$$B_y(t) = \sum_{k=0}^{n} C_{n-k} y_{n-k} s^k \quad (s = 1 - t).$$

In evaluating $B_y(t)$ by Horner's method, the coefficients C_k must be evaluated on the fly. This is done by the inductive process

$$C_0 = 1$$

$$C_{k+1} = C_k \cdot t \cdot \frac{n-k}{k+1}.$$

We run Horner's algorithm with variables P (the current value of the polynomial), k (an index), and C (equal to C_k at all times). We start with $P = y_0$, $k = 1$, and $C = C_1 = nt$. Then the appropriate variant of Horner's algorithm repeats n times, setting in each loop

$$P := P \cdot s + C \cdot y_k$$

$$C := C \cdot \frac{n-k}{k+1} \cdot t$$

$$k := k + 1,$$

and ends with $P = B_y(t)$. It seems plausible that this procedure will be evaluated often, and so efficiency is a major concern. For this reason, in the following code, no dictionary is used. This is one of the more complicated code segments in which I do this (note the heavy use of index to retrieve the values of y, n, s, and t). This is especially reasonable because they remain constant throughout the calculation once stored.

```
% t y=[ y0 y1 ... yn ]
/bernstein {  % t y
    % constants y n t s=1-t
    % variables k C P
    dup length          % t y n+1
    1 sub               % t y n
    3 -1 roll 1         % y n t 1
```

```
1 index sub            % y n t s
 % constants in place
1                      % y n t s k
3 index 3 index mul    % y n t s k C=nt
5 index 0 get          % y n t s k C P=y0
5 index {              % y n t s k C P
   % P -> P* = s.P + C.y[k]
   % C -> C* = C.t.(n-k)/(k+1)
   % k -> k* = k+1
   3 index mul         % y n t s k C P.s
   1 index             % y n t s k C P.s C
   7 index             % y n t s k C P.s C y
   4 index get mul add % y n t s k C P.s+C.y[k]=new P
   3 1 roll            % y n t s P* k C
   5 index             % y n t s P* k C n
   2 index sub mul     % y n t s P* k C.(n-k)
   1 index 1 add div   % y n t s P* k C.(n-k)/(k+1)
   4 index mul         % y n t s P* k C*
   3 1 roll 1 add      % y n t s C* P* k*
   3 1 roll            % y n t s k* C* P*
 } repeat
 7 1 roll 6 { pop } repeat
} def
```

A6.5   CODE

The file `horner.inc` has one procedure `horner` with arguments x and a that returns $a(x)$. The file `bernstein.inc` has a procedure `bernstein` with arguments t and y that returns $B_y(t)$.

APPENDIX 7

Importing PostScript files

Very often you want to import one PostScript file into another. The one you want to import will quite possibly have been produced by another program and may be a more or less generic PostScript file, so you have to be prepared for almost anything. You have to **encapsulate** the imported file so that it does not upset the environment into which it is imported.

A special case of this is one of the most vexing tasks among all those a professional mathematician encounters – that of putting high quality TEX labels into a mathematical diagram. The most general task of this nature can indeed be daunting, but the exact one described here need not be.

I'll explain what to do by a simple example and then add remarks on fancier or more difficult variations.

A7.1   LABELING A GRAPH

Let's suppose you have created the graph of a parabola:

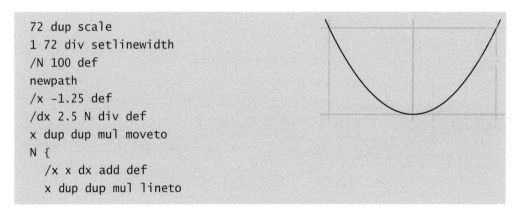

```
72 dup scale
1 72 div setlinewidth
/N 100 def
newpath
/x -1.25 def
/dx 2.5 N div def
x dup dup mul moveto
N {
    /x x dx add def
    x dup dup mul lineto
```

294

```
} repeat
```

```
stroke
```

Now you want to add a label to it, and so it becomes

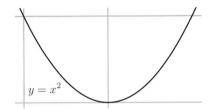

You could produce your own label in PostScript, but getting the fonts to look right, and getting the spacing right in mathematical text – for example, the superscript in this case – is hard and better left to some other program. I have used Donald Knuth's program TeX here to make up the label and then dvips, a program written by John Hobby (once a graduate student of Knuth's), to produce a PostScript file from the TeX output. I'll say more about this process later even though it is not really a PostScript matter. The important thing is that in the end I get a file called, for instance, yx2.eps that contains the PostScript code to write the text "$y = x^2$." The basic idea is now simple: to include a line (yx2.eps) run in the PostScript file containing the parabola. There are two additional things to do, however: (1) take into account the different coordinate systems in parabola drawing and in the label file; (2) isolate effects of the program in the label file.

Dealing with the coordinates is simple – at least in normal circumstances. Dealing with the second will usually involve only annulling the effect of a possible showpage in the imported file if it, at least, is as well behaved as it ought to be. My new file, producing the label and the parabola, now looks like this:

```
gsave
% ... parabola drawing as listed above ...
grestore

gsave
10 10 translate
-290 -695 translate
save /SavedState exch def
/showpage {} def
(yx2.eps) run
```

```
SavedState restore
grestore
```

You might not have encountered `save` and `restore` before. The command `save` does two things: (i) it puts a record of the complete current environment on the stack, and (ii) it saves the current graphics state as `gsave` does. Thus, `save /Saved-State exch def` (*why not* `/SavedState save def`?) defines `SavedState` to be the current state. The command `restore` has one argument, a complete state in the format produced by `save`. The point of using them here, among other things, is to take care of possible `showpage` commands in the imported file but then to bring back the normal definition of `showpage` after the file is read. Unless `showpage` is disabled, an importing program might try to turn a page after reading each import. This technique is especially important in versions 8.0 and later of Ghostscript, *which seem to add implicitly a* `showpage` *command to EPS files it reads* even if there isn't one there originally.

Another and perhaps simpler way to turn off `showpage` is to use a temporary dictionary:

```
1 dict begin
/showpage {} def
(yx2.eps) run
end
```

But `save ... restore` is a more flexible technique. Now all you have to know is where the mysterious numbers 290 and 695 come from. But that's easy. The file produced by `dvips` is an **encapsulated PostScript file** (or EPS file), which means, among other things, that it contains near the beginning a line

```
%%BoundingBox: 290 695 321 708
```

to advertise to applications that want to use it (and that includes us!) what the boundaries of its drawing area – its **bounding box** – are. These are the coordinates of the lower left and upper right corners of that box. The image that TEX and `dvips` produced, in other words, was intended to be placed on a page as on the left with the label sitting inside its bounding box as in the close-up on the right:

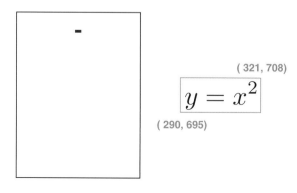

(321, 708)

$$y = x^2$$

(290, 695)

The line -290 -695 translate therefore sets the lower left corner of the imported image at the origin of the figure it is being imported to. So dealing with different coordinate systems is very simple, actually. All you need to do is figure out the bounding box of the imported file. The line 10 10 translate placed before the importation thus has the effect of locating this corner at $(10, 10)$ in the importing file. You might very well want to do some other transformations to that imported file. Suppose that in addition you want to scale the import in place, for example, make it

```
10 10 translate
4 4 scale
-290 -695 translate
```

Order is important here as it always is in coordinate changes.

Deciding where to place the imported file may not be straightforward and usually requires some fiddling around. Most Ghostscript viewers track the location of the mouse in default coordinates to help you out in this task. The viewer I use, for example, records those coordinates – here $(12, 307)$ – in the upper left corner. This is often extremely useful. To take advantage of this feature, you *must* restore default coordinates before importing files.

What I have said so far will handle most imports, but sometimes a somewhat more robust technique is required. The principal new feature is to restore the default graphics environment before importation. Also, messy stack handling by the imported files has to be allowed for.

```
/BeginImport {
    % save the current state
    save /SavedState exch def
    % save the sizes of two stacks
    count /OpStackSize exch def
    /DictStackSize countdictstack def
    % turn off showpage
    /showpage {} def
    % set up default graphics state
    0 setgray 0 setlinecap
    1 setlinewidth 0 setlinejoin
    10 setmiterlimit [] 0 setdash newpath
    /languagelevel where
    {pop languagelevel 1 ne
```

```
        {false setstrokeadjust false setoverprint} if
   } if
} bind def

/EndImport {
  count OpStackSize sub
  dup 0 gt { {pop} repeat} {pop} ifelse
  countdictstack DictStackSize sub
  dup 0 gt { {end} repeat} {pop} ifelse
  SavedState restore
} bind def
```

followed by code like this:

```
BeginImport

% --- import stuff here

EndImport
```

<h2>A7.2 IMPORTING T_EX TEXT</h2>

In this section I want to say more specifically about how to import text produced by TEX, although it is not directly connected with PostScript. Those not familiar with TEX may skip it. I also restrict myself to plain TEX, since the unfortunately more popular variant **Latex** does all kinds of extra formatting I don't want to deal with.

Generally, you will want to import small fragments of mathematics. The only serious requirement for doing this correctly is that, at least in plain TEX, you must put the line \nopagenumbers at the start of your file. This means that there will be no secret writing on your output. Then TEX your file as usual. To produce PostScript output, you should run dvips on the corresponding .dvi file, but with the -E option:

```
dvips -E x.dvi -o x.eps
```

if x.tex is your TEX file. This produces an EPS file with the bounding box data written into it. If you use dvips without the -E option, you will get a full page with your TEX on it, whereas the -E option will produce a bounding box just covering the text you want.

There is one problem with this procedure. The file produced will be quite large. For example, the file yx2.eps I used as an example in the first section was about $17,000$ bytes long! The reason is that dvips puts in several header files and font definitions, and they take up space. These can be amalgamated, but I'm not going to say anything about that here except to raise the issue. Anyway, in the modern world $17,000$ bytes is not all that large, although if you have many labels the redundant code will add up.

If you are using TEX, then you will probably not be able to avoid meeting the PostScript files produced by dvips. It's also likely that sooner or later you'll want to modify one of these by hand, and so I'll say something here about their structure. In fact, I'll look at the one responsible for the label "$y = x^2$" dealt with earlier in this appendix. It starts off with something like

```
%!PS-Adobe-2.0 EPSF-2.0
%%Creator:  dvips(k) 5.86 Copyright 1999 Radical Eye Software
%%Title:  yx2.dvi
%%BoundingBox:  290 695 321 708
%%DocumentFonts:  CMMI10 CMR10 CMR7
%%EndComments
%DVIPSWebPage:  (www.radicaleye.com)
%DVIPSCommandLine:  dvips -o yx2.eps -E yx2.dvi
%DVIPSParameters:  dpi=600, compressed
%DVIPSSource:  TeX output 2003.03.12:1926
```

These are all comment lines. The initial one declares that this file conforms to the conventions for document structure mentioned in an Appendix 3. It also declares that this is an encapsulated PostScript file. The comments beginning with %% are part of the document structure (see Appendix 3). Next comes two sections like this:

```
%%BeginProcSet:  texc.pro
%!
/TeXDict 300 dict def TeXDict begin/N{def}def/B{bind def}N/S{exch}N/X{S
...
end
%%EndProcSet
%%BeginProcSet:  texps.pro
...
%%EndProcSet
```

This just contains two fragments of PostScript code that were imported by dvips to set up its own dictionaries. Next comes

```
%%BeginFont:  CMR7
%!PS-AdobeFont-1.1:  CMR7 1.0
%%CreationDate:  1991 Aug 20 16:39:21
%Copyright (C) 1997 American Mathematical Society.   All Rights Reserved.
...
%%EndFont
```

which just defines a font to be used. More fonts follow. The structure of fonts is a very specialized part of PostScript and is a world pretty much isolated from the rest of the language. Among other things, font files usually contain large chunks of almost indecipherable code.

Finally, the last few lines of the file are

```
TeXDict begin 40258437 52099154 1000 600 600 (yx2.dvi)
@start /Fa 205[33 50[{}1 58.1154 /CMR7 rf /Fb 194[65
61[{}1 83.022 /CMR10 rf /Fc 134[41 47 120[{}2 83.022
/CMMI10 rf end
%%EndProlog
%%BeginSetup
TeXDict begin
%%EndSetup
1 0 bop 1830 183 a Fc(y)26 b Fb(=)d Fc(x)2032 148 y Fa(2)p
eop
%%Trailer
end
userdict /end-hook known{end-hook}if
%%EOF
```

This is the real meat of the file. Some abbreviations are defined, and then the actual typesetting is done in the single line

```
1 0 bop 1830 183 a Fc(y)26 b Fb(=)d Fc(x)2032 148 y Fa(2)p
```

This doesn't make much sense immediately, but actually it's pretty simple. The program dvips abbreviates font changes for efficiency as well as other commonly used PostScript commands. It defines bop to begin pages. So all that's going on here is that characters from different fonts are being placed on the page, and the characters being placed are those in the string "$y = x^2$." As with a modern army at war, then, almost all the effort goes into logistics!

A7.3 FANCY WORK

You can in fact apply any 2D transformation to the image produced by any PostScript code – for example, that in a file to be imported – without modifying the code itself, at least in most circumstances. Why would you want to do this? Well, this is one of those things about which it is said, "If you have to ask. . . ." Any serious book on mysticism will tell you that the highest levels of experience are not accessible to everyone.

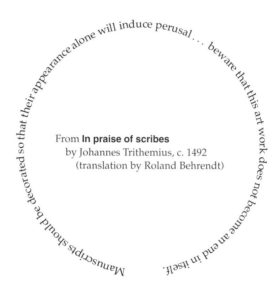

From **In praise of scribes**
by Johannes Trithemius, c. 1492
(translation by Roland Behrendt)

The techniques involved here are much less trivial than others in this book. The difficult point here is that you do not want to examine the internals of the code of the transformed image, except perhaps taking into account its size, as specified by its bounding box. This means that *some of the very basic PostScript drawing commands must be modified* instead. There are several approaches to this problem; the basic choice is whether to modify the paths in the imported code as they are laid down or as they are drawn. In the first scheme the operators moveto etc. are redefined, and in the second stroke etc. I'll choose here the second method. It requires quite a bit less programming, is slightly more efficient, and is also somewhat more flexible.

There are only a few operators involved in actually realizing a path in PostScript as opposed to building it. The obvious ones are stroke, fill, and clip. In addition there is the operator show, which essentially fills in the path made up by a string of characters in whatever the current font is. There are others – principally those concerned with **user paths**, which I have not talked about – but I'll ignore them here. In redefining the drawing operators, it must be kept in mind that the coordinate system may be changed continually in the segment of code to be transformed. For this reason, a base coordinate system in which the transformation is to be applied must be fixed.

One concern that has to be taken into account is that transforming a path will mean transforming the pieces of that path, which may be line and curve segments. Doing this presupposes that these pieces are small enough that the transformation is well approximated by an affine transformation on them. This may not be valid for the original path, and so we must allow for path subdivision.

Yet another concern is that we want to be able to escape from our redefinitions since we might want to keep on drawing normally after we have drawn the transformed imported file. This will be handled by including the redefinitions in a special dictionary that can be pushed and popped on and off the dictionary stack with `begin` and `end`.

All these things are handled in a package `transform.inc`. It has procedures

integer	set-sd	sets subdivision depth
procedure	set-transform	defines the transform
PS matrix	set-base	sets coordinates for the transform to be applied in
	subdivide	subdivides the current path
	path-transform	applies the transform at hand to the current path

Just below is the part of the code that manages the final drawing. The file `trithemius.eps` contains the text *Manuscripts ... itself* all laid out in a line. The file `trithemius2.eps` contains the text to be placed inside the circle just as is. The bounding box data are taken from the file `trithemius.eps` to be used to transform it correctly.

As for the transform, it knows nothing of the contents of the file it is transforming and simply wraps a rectangle around a circle starting at an angle A_0 and ending at an angle A_1, which in this case have been chosen at $-90° \pm 7°$.

```
% define the base coordinate system
/B matrix currentmatrix def
(transform.inc) run
 % --- first the outer text ---
 % set up the transform
 % the bounding box of the imported file
/llx 81 def
/lly 713 def
/urx 389 def
/ury 719 def
```

```
% the dimensions of the box we are displaying
/boxwidth urx llx sub def
/boxheight ury lly sub def
 % the angles where the circular text starts and ends
/A0 270 7 sub def
/A1 -90 7 add def
 % radius of the circular text
/radius 100 def
 % proportion of a full circle the text takes up
/factor A0 A1 sub 360 div def
 % the length it will take up on the circle
/truelength factor 3.1416 mul 2 mul radius mul def
 % the transform itself
/f { 1 dict begin
  /y exch lly sub def
  /x exch llx sub def
  /T 1 x boxwidth div sub A0 A1 sub mul A1 add def
  /R 100 boxheight 2 div sub y truelength boxwidth div mul add def
  T cos R mul T sin R mul
end } def

gsave
transformdict begin
  /f load set-transform
   % set the base coordinate system
  B set-base
  1 dict begin
  /showpage{} def
  (trithemius.eps) run
  end      % the temporary dictionary
end       % transformdict
grestore
 % --- now the inner text ---
0.92 dup scale
-91 -683 translate
58 98 translate

1 dict begin
/showpage{} def
(trithemius2.eps)  run
end
```

REFERENCES

1. Adobe Systems, **Adobe Type 1 Font Format**, 1990. This is the original edition of the definitive document. Newer versions are available on line.

2. Johannes Trithemius, **In Praise of Scribes**, translated into English from **De Laude Scriptorum** by Roland Behrendt, Coronado Press, 1974.

Epilogue

The association between mathematics and graphics is ancient. Indeed, as the scholar of Greek mathematics T. L. Heath has observed, the Greek root of the word "graphics" seems to mean in places "to prove." But the association is far older than even the civilization of classical Greece. One of the very oldest mathematical documents we now possess is YBC 7289, a tablet dating from about 1800 B.C. and now found in the Yale Babylonian Collection. The three numbers written on the tablet (in base 60 notation) express $1/2$, an approximation of $\sqrt{2}$ to about eight decimal figures, and the corresponding product $\sqrt{2}/2$. The diagram on it is plausibly a part of a simple geometric proof of Pythagoras's theorem for isosceles right triangles closely related to the well-known figure associated with the discussion of this result in Plato's *Meno*. This extraordinary object therefore seems to show that the person who made it knew both that the ratio of the diagonal to the side of a square is a real number whose square is two and why it is so. I like to think that this diagram tells us that the association of logical reasoning with mathematics originated with deductions from figures, although of course any train of real evidence of how mathematical reasoning came about to us is beyond recovery.

The quality of mathematical reasoning made an extraordinary leap among the Greeks, notably with the appearance of the *Elements* of Euclid. Figures were still required, of course, and one small but significant advance was made by linking text and figures with labels. Of course the connection between pictures and reasoning has continued for the entire history of mathematics, although with varying importance. It reached a low point during the eighteenth century probably because, from the earliest days of modern mathematical analysis, it was realized that pictures are

> On ne trouvera point de Figures dans cet Ouvrage. Les méthodes que j'y expose ne demandent ni conſtructions, ni raiſonnemens géométriques ou méchaniques, mais ſeulement des opérations algébriques, aſſujetties à une marche réguliere & uniforme. Ceux qui aiment l'Analyſe, verront avec plaiſir la Méchanique en devenir une nouvelle branche, & me ſauront gré d'en avoir étendu ainſi le domaine.

inadequate to deal with the complexities of the subject and can be seriously deceptive. The tone was struck most forcefully in Lagrange's famous boast in the preface to his *Analytical mechanics* that "One will not find any figures in this work. The methods which I explain in it require neither constructions nor geometrical nor mechanical reasoning, but only algebraic operations." Such an attitude towards illustration in mathematics continued to be sounded, and with some justification, through the entire nineteenth century and into the twentieth. During much of that time, it was often amateurs who came up with the best graphics in mathematics, although there were notable exceptions. Technology was poor. Even through the nineteenth century technical illustrations were often done with woodcuts. To compensate for this, labor was fairly cheap during much of the century, but eventually costs overtook convenience, and the quality of mathematical illustration went down. Cost presumably explains why we had otherwise fine books on geometry in the twentieth century notoriously lacking illustrations such as Julian Coolidge's **History of Geometrical Methods**. (In speaking of the origins of perspective drawing he writes "Alberti does not explain himself very clearly, and we may say that . . . the easiest way to understand those early works on perspective is to study the pictures, not to read the text." But Coolidge has no pictures!) Later in that century, computers restored to mathematics the potential of being associated with great graphics – one that has even been occasionally realized. But although computers have made it possible to publish great pictures, technology alone does not guarantee that the worth of a picture reflects how much work went into it. It is important to think carefully about exactly what one wants. Pictures should be drawn with as much care as paragraphs are written.

Edward Tufte, now retired from a career as Professor at Yale University, has written several books on what he calls "information graphics." They are very beautiful books – produced by his own press, which was founded expressly for the purpose – and indeed these books illustrate that the medium really is often the message. Nonetheless, there are a few more or less practical rules to be extracted from them that could be of value to anybody who wants to explain something by visual means. Much Tufte's effort goes into the display of *data* – weather patterns, train schedules, the attrition in Napoleon's army on the long, cold, winter road to Moscow – but many of his ideas apply to the task of mapping out a train of logic. I'll give an example here.

I will do this by comparing what I am embarrassed to call the traditional mode of mathematical exposition with one adapted from Tufte's suggestions.

I am going to look at the classical result, known to the ancient Greeks, that the golden ratio is irrational – that is, cannot be expressed as a ratio of whole numbers. This is of course one of the oldest mathematical discoveries and perhaps the first truly astonishing one. There is much discussion in the literature – almost entirely speculative, of course – as to how incommensurability was first found. One common and reasonable speculation is that it was arrived at by geometric reasoning – not that geometry provided at first a completely rigourous proof in view of the then primitive state of Greek mathematics, but that it at least provided a convincing chain of reasoning of some kind leading to the result.

The golden ratio is also the ratio between the side and diagonal of a regular pentagon, and what we will actually demonstrate is that *the side and diagonal of a regular pentagon are incommensurable.* Very briefly, the idea of the argument used here is to see that if the side and diagonal are both multiples of an interval ϵ, then so are the side and diagonal of the smaller pentagon at the center of the five-sided star whose vertices are those of the original pentagon. Recursion leads to a contradiction.

I will begin by quoting a very traditional approach to this problem from a 1945 paper by Kurt von Fritz on the discovery of incommensurability. Keep in mind throughout what is to follow that the point, as von Fritz says, is not merely to prove the result but to make it "almost apparent at first sight." Here is my copy of the figure drawn by von Fritz:

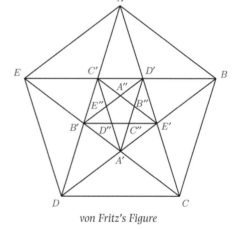

von Fritz's Figure

Here is what he writes:

> ...the diameters of the pentagon form a new regular pentagon in the center,...the diameters of this smaller pentagon will again form a regular pentagon, and so on in an infinite process. It is...easy to see that in the pentagons produced in this way $AE = AB'$ and $B'D = B'E'$ and therefore $AD - AE = B'E'$, and likewise $AE = ED' = EA'$ and $B'E' = B'D = B'E$ and therefore $AE - B'E' = B'A'$, and so forth ad infinitum, or, in other words, that the difference between the diameter and the side of the greater pentagon is equal to the diameter of the smaller pentagon, and the

difference between the side of the greater pentagon and the diameter of the smaller pentagon is equal to the side of the smaller pentagon, and again the difference between the diameter of the smaller pentagon and its side is equal to to the diameter of the next smaller pentagon and so forth in infinitum. Since ever new regular pentagons are produced by the diameters it is then evident that the process of mutual subtraction will go on forever, and therefore no greatest common measure of the diameter and the side of the regular pentagon can be found.

There is nothing wrong in the logic of this treatment, although it does stumble around somewhat. What we are interested in right now, however, is how the argument relates to the figure. The answer, I think it is fair to say, is "not well." Reading the original article is even more difficult than apparent here because, as often happens, the text and figure are on separate pages. What I claim is that *von Fritz stumbles precisely because he is trying to put in words what could have been far better put in pictures*. His one figure is not really used in a serious way and essentially does no more than make the argument unambiguous. I also think it is fair to say that von Fritz is far from making the result apparent at first sight. Contrary to what he wants to do, he is preaching to the converted. Of course one might object that a paper over 50 years old cannot be held completely responsible for its graphics, but actually von Fritz does better than many more recent authors.

Let's see what help Tufte might be able to offer. The first step is to decide to take the graphics more seriously – to make the graphics the main part of the narrative rather than subsidiary to it. The next step is to integrate text and graphics better; this is von Fritz's major failing because, in reading his argument, you are constantly forced to go back to the diagram, relocate yourself there, and so on. A third is to determine which elements of the illustrations are important and then to emphasize them. In von Fritz's figure there are only the labeling of the vertices to orient the reader. But in fact the entities involved are not really the vertices at all but instead various edges and subregions of the pentagon.

Perhaps the most succinct application of Tufte's principles is found in Chapter 4 of **Visual Explanations**, "The Smallest Effective Difference." It opens with a diagram of the ear taken from the **Random House Dictionary of the English Language**, which has a remarkable resemblance to von Fritz's diagram! Tufte redraws it to make it clearer by carrying through the following ideas:

■ Tone down the secondary elements of a picture to reduce visual clutter, to clarify the primary elements of the figure, and also to eliminate unwanted visual interactions. Tufte calls this **layering** the figure to produce a visual hierarchy.

■ Reduce discontinuity in the exposition – that is, replace coding labels in the figure by locally useful information in the figure itself. The general principle is to integrate text and graphics. One point is that unnecessary eye movements are fatal to easy comprehension.

■ Produce emphasis by using the smallest possible effective distinctions. In practice this often, but not always, means replacing bold, strongly contrasting colors by quieter shades. This is perhaps the hardest of all sins to avoid, for it is often extremely tempting for the beginner to introduce strong colors whenever possible. In Web graphics, this does often work well.

To this list might be added a few ideas from elsewhere in Tufte's books:

■ Eliminate parts of the figure that do not actually add to its content. If they do not add to it then they will subtract from it.

■ Use what Tufte calls **small multiples**, numerous repetitions of a single figure with slight variations. Human perception is sharp in making comparisons. Another way to put this, particularly in trying to track a logical argument visually, is to say that the sequence should have **continuity** in the dramatic sense.

■ Make the graphics itself carry the tale as much as possible. Where text is necessary, eliminate one major source of annoyance by placing related text and graphics close to each other.

All of these are nearly self-evident principles, and if the use of graphics in mathematics were more sophisticated than it is now one might consider this an objection to Tufte's books.

Here now is the argument I have made up in an attempt to apply these principles:

 The basic fact is that in a regular pentagon a diagonal and the side opposite to it are parallel. This property in some sense characterizes the regular pentagon.

 As a consequence, the shaded region shown at left is a parallelogram having equal sides throughout (a rhombus).

 Assume now that the side s and diagonal d are commensurable, which is to say that they are both multiples of a common interval ϵ.

Then $d - s$ is also a multiple of the interval ϵ.

And so is the interval we get in the middle of the diagonal, which has length $d - 2(d - s) = 2s - d$.

But this interval is the side of the smaller pentagon at the center of the star we get by drawing all diagonals.

The figure emphasized in the diagram to the left is a parallelogram since opposite sides are parallel to the same side of the pentagon. Therefore, the quantity $d - s$ is the diagonal of the smaller pentagon.

Consequently, under the assumption that the diagonal and side of a pentagon are multiples of a common interval ϵ, we deduce that the side and diagonal of the smaller pentagon inscribed in it are as well.

We can reason in the same way about the pentagon in its interior in turn, and so on. The interval ϵ will divide all the sides and diagonals of the infinite series of pentagons we get. But eventually the sides of those pentagons will be smaller than ϵ, which is a contradiction.

I imagine that some readers will find my argument distasteful. I am, however, in good company. It is no less than J. E. Littlewood, who points out (p. 54 of the *Miscellany*) that

> A heavy warning used to be given that pictures are not rigourous; this has never had its bluff called and has permanently frightened its victims into playing for safety.

Littlewood's remark accompanies an elegant "picture proof" of a lemma of Landau's.

REFERENCES

1. Bill Casselman, A review of **Visual Explanations** by E. Tufte, *Notices of the AMS* **46**, January 1999, pp. 43–46. Much of this epilogue was taken from this.

2. K. von Fritz, "The discovery of incommensurability by Hppasus of Metapontium," *Annals of Mathematics* (1945), pp. 242–264.

3. T. L. Heath, **A History of Greek Mathematics volume I**, Dover, 1981. The discussion of "drawing" and "proof" is in the extensive footnote on p. 203. This question has been taken up more recently by Wilbur Knorr in Section III.II of **The Evolution of Euclid's Elements**, and again in Section 2.2 of **The Shaping of Deduction**, an exhaustive investigation of Greek mathematical diagrams by the classical scholar Reviel Netz. Both of these later discussions are very interesting – especially for those like me who believe on a priori grounds that mathematical proofs originated with diagrams, where visualization led inevitably to reasoning.

4. J. E. Littlewood, **Littlewood's Miscellany**, Cambridge University Press, 1988.

5. E. Tufte, **The Visual Display of Quantitative Information**, Graphics Press, Cheshire, Connecticut, 1983.

6. —, **Envisioning Information**, Graphics Press, 1990.

7. —, **Visual Explanations**, Graphics Press, 1997.

Index